Multirate and Wavelet
Signal Processing

Wavelet Analysis and Its Applications

The subject of wavelet analysis has recently drawn a great deal of attention from mathematical scientists in various disciplines. It is creating a common link between mathematicians, physicists, and electrical engineers. This book series will consist of both monographs and edited volumes on the theory and applications of this rapidly developing subject. Its objective is to meet the needs of academic, industrial, and governmental researchers, as well as to provide instructional material for teaching at both the undergraduate and graduate levels.

Among the attractive features of wavelet analysis is the computational aspect of the subject. In particular, computation of the discrete wavelet transform can be accomplished by filter bank algorithms in subband coding. This eighth volume of the series is an elementary treatise of the subject of multirate, including the detailed discussion of filter banks and their lattice structures, as well as an application of multirate to wavelet implementation.

The series editor would like to congratulate the author for an insightful presentation of an important area in wavelet analysis.

This is a volume in
WAVELET ANALYSIS AND ITS APPLICATIONS

CHARLES K. CHUI, SERIES EDITOR

A list of titles in this series appears at the end of this volume.

Multirate and Wavelet Signal Processing

Bruce W. Suter

Air Force Institute of Technology (AFIT/ENG)
Department of Electrical and Computer Engineering
Wright-Patterson Air Force Base
Ohio

ACADEMIC PRESS
San Diego   London   Boston
New York   Sydney   Tokyo   Toronto

Copyright © 1998 by Academic Press

ACADEMIC PRESS
525 B Street, Suite 1900, San Diego, CA 92101-4495, USA
1300 Boylston Street, Chestnut Hill, MA 02167, USA
http://www.apnet.com

ACADEMIC PRESS LIMITED
24–28 Oval Road, London NW1 7DX, UK
http://www.hbuk.co.uk/ap/

Library of Congress Cataloging-in-Publication Data

Suter, Bruce W.
    Multirate and wavelet signal processing / Bruce W. Suter.
      p.   cm. — (Wavelet analysis and its applications ; v. 8)
    Includes bibliographical references and index.
    ISBN 0-12-677560-5
    1. Signal processing—Digital techniques—Mathematics.
2. Electrical filters—Mathematics. 3. Wavelets (Mathematics).
I. Title. II. Series.
TK5103.7.S94   1997
621.382'2—dc21
                              97-30328
                              CIP

Printed in the United States of America
97 98 99 00 01 IC 9 8 7 6 5 4 3 2 1

This book was only possible through the patience, understanding, encouragement, and support of my wife Debbie.
As such, I dedicate this book to her.

Contents

Preface

The field of multirate and wavelet signal processing finds applications in speech and image compression, the digital audio and digital video industries, adaptive signal processing, and in many other applications.

The utilization of multirate techniques is becoming an indispensable tool of the electrical engineering profession. This point can be illustrated in three ways. First, if a performance specification is controlling the design of a particular system, that is, the performance specification exceeds the current state-of-art, then by converting the system to a multirate system, the overall system specification can be met with slower components. Secondly, if the dollar cost specification is controlling the design of a particular system, that is, the design of a competitive commercial system where bottom line cost is most important, then by converting the system to a multirate system, the overall system cost will be reduced through the utilization of slower, cheaper devices. Thirdly, if power consumption is controlling the design of a particular system, that is, the design of a hand-held system powered by a couple AA batteries, or possibly a satellite system, then by converting the system to a multirate system will reduce power consumption through the utilization of devices with slower switching speed, and as a result, lower power dissipation.

Wavelet transforms are closely related to filter banks. As such, a background in filter banks will make it easier for the reader to understand, design, and implement wavelet transforms.

Many of the most important applications, such as video compression, and many challenging research problems are in the area of multidimensional multirate. As such, multidimensional multirate is integrated throughout the book.

The focus of this book is to present a sound theoretical foundation by emphasizing the general principles of multirate. This book is self-contained for readers who have some prior exposure to linear algebra (at the level of Horn and Johnson's *Matrix Analysis*) and multidimensional signal processing (at the level of Lim's *Two-Dimensional Signal and Image Processing* or Dudgeon and Mersereau's *Multidimensional Digital Signal Processing*).

xi

Moreover, this text will bring the reader to a point where he/she can read, understand, and appreciate the vast multirate literature.

The organization of this book is as follows. The first two chapters are devoted to basic multirate ideas including decimators, expanders, polyphase notation, etc. This presentation is first given for one-dimensional signals in Chapter 1 and then generalized to multidimensional signals in Chapter 2. The next two chapters deal with filter banks. Chapter 3 presents the theory of filter banks for both one-dimensional and multidimensional signals. Chapter 4 deals with lattice structures, an efficient implementation strategy for filter banks. Chapter 5 highlights an important application of multirate — the implementation of wavelets.

I would also like to take this opportunity to thank Professor Charles Chui for his enthusiasm about this project and for including this text in his distinguished wavelet series. The following people have provided very useful feedback during the writing of this book. They include: Bill Cowan, Tom Foltz, Jerry Gerace, Ying Huang, You Jang, Matt Kabrisky, Mark Oxley, Robert Parks, Juan Vasquez, and Dan Zahirniak.

Fairborn, Ohio   Bruce W. Suter
February 9, 1997

Multirate and Wavelet
Signal Processing

Chapter 1

Multirate Signal Processing

1.1 Introduction

This chapter provides the basic concepts used in the study of multirate and wavelet signal processing. Some of the earliest contributions to the study of the fundamentals of multirate were due to Schafer and Rabiner[40], Meyer and Burrus[32], Oetken *et al.*[37], and Crochiere and Rabiner[10]. The idea of polyphase representation is a key concept throughout the development of this book. This nontrivial idea was first articulated by Bellanger *et al.*[3]. Much more recently, Evangalista[17] carefully examined another important idea – digital comb filters.

Many of the concepts developed in this chapter are also discussed in the other multirate texts by Crochiere and Rabiner[11], Fliege[19], Strang and Nguyen[46] and Vaidyanathan[49].

Section 1.2 presents a framework for multirate and it introduces two important representations for discrete signals. Section 1.3 introduces the basic building blocks. Section 1.4 provides ways to interchange the basic building blocks. Section 1.5 presents a filter bank example.

1.2 Foundations of multirate

First we will examine some sampling considerations and then present some basic transforms for analyzing signals.

1.2.1 Sampling considerations

Multirate is the study of time-varying systems. As such, the sampling rate will change at various points in time in an implementation. This will require us to vary the gain (magnitude) of filters in series with the time-varying building blocks so that the resulting gain is consistent with what one would expect if the sampling interval after the time-varying block had been the original sampled frequency. Towards this end, let us analyze a train of impulses.

Theorem 1.2.1.1. $\sum_{k=-\infty}^{\infty} \delta(t - kT) = \frac{1}{T} \sum_{m=-\infty}^{\infty} \exp(\frac{j2\pi mt}{T})$.

Proof: Let us expand $\sum_{k=-\infty}^{\infty} \delta(t - kT)$ in a Fourier series. So,

$$\sum_{k=-\infty}^{\infty} \delta(t - kT) = \sum_{m=-\infty}^{\infty} a(m) \exp\left(\frac{j2\pi mt}{T}\right)$$

where,

$$a(m) = \frac{1}{T} \int_{-\frac{T}{2}}^{\frac{T}{2}} \left[\sum_{k=-\infty}^{\infty} \delta(t - kT)\right] \exp\left(-\frac{j2\pi mt}{T}\right) dt$$

or equivalently,

$$a(m) = \frac{1}{T} \sum_{k=-\infty}^{\infty} \int_{-\frac{T}{2}}^{\frac{T}{2}} \delta(t - kT) \exp\left(-\frac{j2\pi mt}{T}\right) dt.$$

Let $\tau = t - kT$. Then,

$$a(m) = \frac{1}{T} \sum_{k=-\infty}^{\infty} \int_{-\frac{T}{2}-kT}^{\frac{T}{2}-kT} \delta(\tau) \exp\left(-\frac{j2\pi m(\tau + kT)}{T}\right) d\tau.$$

We recognize this as a sum of integrals with adjoining limits and simplify to

$$a(m) = \frac{1}{T}.$$

Hence,

$$\sum_{k=-\infty}^{\infty} \delta(t - kT) = \frac{1}{T} \sum_{m=-\infty}^{\infty} \exp\left(\frac{j2\pi mt}{T}\right).$$

∎

Let $\mathcal{F}$ denote the Fourier transform. So that if $x(t)$ is a signal, then

$$\mathcal{F}[x(t)] = \int_{-\infty}^{\infty} x(t) \exp(-j2\pi ft) dt.$$

Let us examine the Fourier transform of an impulse train.

Theorem 1.2.1.2. $\mathcal{F}\left[\sum_{k=-\infty}^{\infty} \delta(t - kT)\right] = \frac{1}{T} \sum_{m=-\infty}^{\infty} \delta(f - \frac{m}{T})$.

Proof: From the previous theorem,

$$\sum_{k=-\infty}^{\infty} \delta(t - kT) = \frac{1}{T} \sum_{m=-\infty}^{\infty} \exp\left(\frac{j2\pi mt}{T}\right).$$

Therefore,

$$
\begin{aligned}
\mathcal{F}\left[\sum_{k=-\infty}^{\infty} \delta(t - kT)\right] &= \mathcal{F}\left[\frac{1}{T}\sum_{m=-\infty}^{\infty} \exp(\tfrac{j2\pi mt}{T})\right] \\
&= \frac{1}{T}\int_{-\infty}^{\infty}\sum_{m=-\infty}^{\infty}\exp(\tfrac{j2\pi mt}{T})\exp(-j2\pi tf)dt \\
&= \frac{1}{T}\sum_{m=-\infty}^{\infty}\int_{-\infty}^{\infty}\exp\left[-j2\pi t\left(f - \tfrac{m}{T}\right)\right]dt \\
&= \frac{1}{T}\sum_{m=-\infty}^{\infty}\delta(f - \tfrac{m}{T}).
\end{aligned}
$$

∎

Now, performing the Fourier transform of a sum of the product of the input $x(t)$ and Dirac delta functions, which can be expressed as the convolution of the corresponding functions, produces

$$
\begin{aligned}
\mathcal{F}\left[\sum_{k=-\infty}^{\infty} x(t)\delta(t - kT)\right] &= \int_{-\infty}^{\infty} X(f - f')\frac{1}{T}\sum_{k=-\infty}^{\infty}\delta\left(f' - \tfrac{k}{T}\right)df' \\
&= \frac{1}{T}\sum_{k=-\infty}^{\infty}\int_{-\infty}^{\infty} X(f - f')\delta\left(f' - \tfrac{k}{T}\right)df' \\
&= \frac{1}{T}\sum_{k=-\infty}^{\infty} X\left(f - \tfrac{k}{T}\right).
\end{aligned}
$$

At this point, we will interpret these results for linear time-varying systems. If the sampling interval is increased by an integer factor of M, where $M > 1$, then the magnitude of the Fourier transform will need to be decreased by a factor of $\frac{1}{M}$ to reconstruct the original system, that is

$$
\mathcal{F}\left[\sum_{k=-\infty}^{\infty} x(t)\delta(t - kMT)\right] = \frac{1}{MT}\sum_{k=-\infty}^{\infty} X\left(f - \frac{k}{MT}\right).
$$

Similarly, if the sampling interval is decreased by an integer factor of L, where $L > 1$, then the magnitude of the Fourier transform will need to be increased by a factor of L to reconstruct the original signal, that is

$$
\mathcal{F}\left[\sum_{k=-\infty}^{\infty} x(t)\delta\left(t - \frac{kT}{L}\right)\right] = \frac{L}{T}\sum_{k=-\infty}^{\infty} X\left(f - \frac{kL}{T}\right).
$$

1.2.2 Sampled signals

For completeness, the definition of z-transforms and the discrete Fourier transforms will be presented. Then, we will present two sampled signal representations: the modulation representation and the polyphase representation. The theory of multirate and wavelet signal processing utilizes both of these representations

1.2.2.1 z-Transforms

Definition 1.2.2.1. *The z-transform of a sequence $x(n)$ is defined by*

$$X(z) = \mathcal{Z}\left[x(n)\right] = \sum_{n=-\infty}^{\infty} x(n)z^{-n}.$$

An important property of z-transforms is the following scaling theorem.

Theorem 1.2.2.1. *If the z-transform of x exists and α is a scalar, then*

$$\mathcal{Z}\left[\alpha^{-n}x(n)\right] = X(\alpha z).$$

Proof: By definition,

$$\mathcal{Z}\left[\alpha^{-n}x(n)\right] = \sum_{n=-\infty}^{\infty} \alpha^{-n}x(n)z^{-n}$$

or equivalently,

$$\mathcal{Z}\left[\alpha^{-n}x(n)\right] = \sum_{n=-\infty}^{\infty} x(n)(\alpha z)^{-n}.$$

Hence,

$$\mathcal{Z}\left[\alpha^{-n}x(n)\right] = X(\alpha z).$$

$\blacksquare$

If the z-transform converges for all z of the form $z = \exp(j\omega)$ for real ω, then the z-transform can be represented as the sum of harmonically related sinusoids, *i.e.*

$$X(\omega) = \sum_{n=-\infty}^{\infty} x(n)\exp(-j\omega n),$$

which is sometimes called the discrete-time Fourier transform.

1.2.2.2 Discrete Fourier transform

Definition 1.2.2.2. *The discrete Fourier transform of a periodic sequence $x(n)$ of length N is given by*

$$X(k) = \sum_{n=0}^{N-1} x(n)\exp\left(-\frac{j2\pi kn}{N}\right); k = 0,\ldots,N-1$$

and the corresponding *inverse discrete Fourier transform* is given by

$$x(n) = \frac{1}{N}\sum_{k=0}^{N-1} X(k)\exp\left(\frac{j2\pi kn}{N}\right); n = 0,\ldots,N-1.$$

Let $W_N = \exp\left(\frac{-j2\pi}{N}\right)$, the (principal) Mth root of unity. Then, the discrete Fourier transform matrix, denoted $\mathbf{W}_N$, is an N x N matrix, defined by

$$[\mathbf{W}_N]_{k,n} = W_N^{kn} = \exp\left(\frac{-j2\pi kn}{N}\right).$$

1.2.2.3 Modulation representation

Definition 1.2.2.3. *Given a sequence $x(n)$ and a positive integer M, then the components of the modulation representation of the z-transform of $x(n)$ are defined $X(zW_M^k)$, $k = 0, 1, \ldots, M - 1$.*

If $M = 2$, then the components of the modulation representation are $X(z)$ and $X(-z)$. The term modulation representation can be most easily visualized in the time domain. Using the scaling theorem of z-transforms, we obtain

$$
\begin{aligned}
\mathcal{Z}^{-1}[X(zW_M^k)] &= \left(W_M^{-k}\right)^n x(n) \\
&= \exp\left(\frac{j2\pi kn}{M}\right) x(n).
\end{aligned}
$$

It is interesting to note that the components of the modulation representation can be combined pairwise to form a real signal, that is

$$
\begin{aligned}
\mathcal{Z}^{-1}[X(zW_M^k) + X(zW_M^{M-k})] &= W_M^{-kn}x(n) + W_M^{-Mn+kn}x(n) \\
&= 2\cos(\tfrac{2\pi kn}{M})x(n).
\end{aligned}
$$

1.2.2.4 Polyphase representation

Definition 1.2.2.4. *Given a sequence $x(n)$ and a positive integer M, then the Type-I polyphase components of $x(n)$ are defined as $x_k(n) = x(Mn+k)$, $k = 0, 1, \ldots, M - 1$.*

If $M = 2$, then $x_0(n)$ would be the even-numbered samples and $x_1(n)$ would be the odd-numbered samples. Now, let us investigate the z-transform of the Type-I polyphase components, that is

$$X(z) = \sum_{n=-\infty}^{\infty} \sum_{k=0}^{M-1} x(Mn + k)\, z^{-(Mn+k)}$$

or equivalently,

$$X(z) = \sum_{k=0}^{M-1} z^{-k} \sum_{n=-\infty}^{\infty} x(Mn + k)(z^M)^{-n}.$$

Let

$$X_k^{(M)}(z) = \sum_{n=-\infty}^{\infty} x(Mn+k)z^{-n} = \sum_{n=-\infty}^{\infty} x_k(n)z^{-n}$$

then $X(z)$ becomes

$$X(z) = \sum_{k=0}^{M-1} z^{-k} X_k^{(M)}(z^M).$$

The success of this decomposition rests on its usage of the well-known Division Theorem for Integers, which says any integer p can be represented as

$$p = Mn + k$$

where, the quotient n and the remainder k are unique integers.

Definition 1.2.2.5. *Given a sequence $x(n)$ and a positive integer M, its Type-II polyphase components of $x(n)$ are given by $x_k(n) = x(Mn + M - 1 - k)$, $k = 0, \ldots, M - 1$.*

The z-transform of $x(n)$ written in the Type-II polyphase representation is given by

$$X(z) = \sum_{n=-\infty}^{\infty} \sum_{k=0}^{M-1} x(Mn + M - 1 - k)\, z^{-(Mn+M-1-k)}$$

or equivalently,

$$X(z) = \sum_{k=0}^{M-1} z^{-(M-1-k)} \sum_{n=-\infty}^{\infty} x(Mn + M - 1 - k)(z^M)^{-n}.$$

Let

$$X_{M-1-k}^{(M)}(z) = \sum_{n=-\infty}^{\infty} x(Mn + M - 1 - k)z^{-n} = \sum_{n=-\infty}^{\infty} x_k(n)z^{-n}$$

then

$$X(z) = \sum_{k=0}^{M-1} z^{-(M-1-k)} X_{M-1-k}^{(M)}(z^M).$$

It is also important to note that Type-II polyphase components are identical to Type-I polyphase components – only the indexing is performed in reverse order.

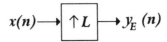

Figure 1.1. L-fold expander.

1.3 Basic building blocks

This section defines and analyzes the two types of basic building blocks used in multirate signal processing. One type of building block deals with changes in sampling rate of the input signal and the other type of building block deals with changes in filter length. We will first discuss two types of building blocks which change the sampling rate of the input signal – *decimators* to reduce it and *expanders* to increase it. Secondly, another type of building block is discussed which changes the length of a given filter – *comb* filters to increase its length. In this way, a comb filter version of a given filter is analogous to an expander acting on an input signal.

1.3.1 Expanders

Definition 1.3.1.1. *Let L be a positive integer. An L-fold expander (also known as an upsampler) applied to an input signal $x(n)$ inserts $L-1$ zeros between adjacent samples of the input signal. For a sampled signal $x(n)$, the output of the expander is given by*

$$y_E(n) = \begin{cases} x\left(\frac{n}{L}\right), & \text{if } n \text{ is a multiple of } L \\ 0, & \text{otherwise .} \end{cases}$$

An L-fold expander is depicted pictorially in Figure 1.1.

Let us consider the following example to gain some intuition into the behavior of expanders. If $L = 2$, then the inputs and outputs of the expander are given by:

$$x(n): \quad x(0) \quad\quad x(1) \quad\quad x(2) \quad\quad x(3) \quad \dots$$
$$y_E(n): \quad x(0) \quad 0 \quad x(1) \quad 0 \quad x(2) \quad 0 \quad x(3) \quad \dots .$$

The operation of expansion is invertible, or in other words, it is possible to recover $x(n)$ from samples of $y_E(n)$. We will now analyze the behavior of the L-fold expander by taking a z-transform of it, that is

$$Y_E(z) = \sum_{n=-\infty}^{\infty} y_E(n) z^{-n},$$

or equivalently,

$$Y_E(z) = \sum_{n \text{ a mult of } L} y_E(n)z^{-n} + \sum_{n \text{ not a mult of } L} y_E(n)z^{-n}.$$

By the definition of the expander, $y_E(n) = 0$ if n is not a multiple of L. Therefore,

$$Y_E(z) = \sum_{n \text{ a mult of } L} y_E(n)z^{-n}.$$

Let $n = kL$. Then,

$$Y_E(z) = \sum_{k=-\infty}^{\infty} y_E(kL)z^{-kL}.$$

By the definition of the L-fold expander, $x(k) = y_E(kL)$. Hence,

$$Y_E(z) = \sum_{k=-\infty}^{\infty} x(k)z^{-kL},$$

or equivalently,

$$Y_E(z) = X(z^L).$$

In the literature, the notation

$$Y_E(z) = X(z)|_{\uparrow L}$$

is occasionally used. If the region of convergence of $Y_E(z)$ includes the points of the form $z = \exp(j\omega)$, where ω is real, then we can replace z with $\exp(j\omega)$ to yield

$$Y_E(\exp(j\omega)) = X(\exp(jL\omega)).$$

By suppressing the exponential and, as such, changing the notation so that $Y_E(\omega)$ denotes $Y_E(\exp(j\omega))$, then we will write

$$Y_E(\omega) = X(L\omega).$$

Therefore, $Y_E(\omega) = X(L\omega)$ means that a given frequency ω_0 in $X(\omega)$ is transformed to a *new* frequency $\frac{\omega_0}{L}$ in $Y_E(\omega)$. As a result of this transformation $L - 1$ unwanted image spectra of the input signal spectra appeared after each of the original spectra at the output of the expander. This suggests the use of a lowpass filter with cutoff frequency $\frac{\pi}{L}$ immediately following the expander as illustrated in Figure 1.2.

This lowpass filter is called an *anti-imaging filter*. Recall from the previous section, if the sampling interval is decreased by a factor of L, then the product of the gains of the expander and the anti-imaging filter must equal L. Since the expander has unity gain, the anti-imaging filter must

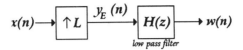

Figure 1.2. Expander with anti-imaging filter.

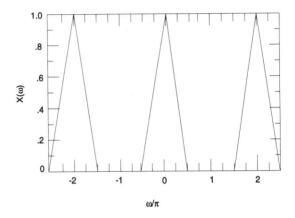

Figure 1.3. Spectrum of input signal.

have a gain of L. As an example, let the spectrum of the input signal $X(\omega)$ be given in Figure 1.3. If the expansion ratio is two, that is $L = 2$, then the spectrum of the expanded signal $Y_E(\omega)$ is given in Figure 1.4. Moreover, the filter given in Figure 1.5 will permit the recovery of $X(\omega)$ from $Y_E(\omega)$.

1.3.2 Decimators

Definition 1.3.2.1. *Let M be a positive integer. An M-fold decimator (also known as a downsampler) applied to the input signal $x(n)$ retains only those samples of the input signal that occur at M-time multiples apart. For a sampled signal $x(n)$, the output of the decimator is given by*

$$y_D(n) = x(Mn).$$

An M-fold decimator is depicted pictorially in Figure 1.6.

Consider the following example to gain some intuition into the operation of decimation. If $M = 2$, then the inputs and outputs of the decimator are given by:

$x(n)$	$x(0)$	$x(1)$	$x(2)$	$x(3)$	$x(4)$	$x(5)$	$x(6)$	$\dots$
$y_D(n)$	$x(0)$		$x(2)$		$x(4)$		$x(6)$	$\dots$.

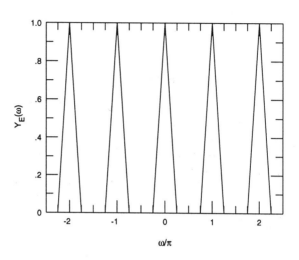

Figure 1.4. Spectrum of expanded signal.

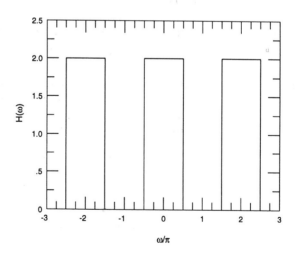

Figure 1.5. Filter to recover input signal.

Figure 1.6. M-fold decimator.

The operation of decimation is not invertible, or in other words, it is not possible to recover $x(n)$ from $y_D(n)$. We will now analyze the behavior of the M-fold decimator by taking a z-transform of it, that is

$$Y_D(z) = \sum_{n=-\infty}^{\infty} y_D(n)z^{-n},$$

or equivalently,

$$Y_D(z) = \sum_{n=-\infty}^{\infty} x(Mn)z^{-n}.$$

At this point we can not make the substitution $k = Mn$ and proceed with a frequency domain expression for $Y_D(z)$. Why? Because $x(n)$ is not zero for noninteger multiples of Mn. So define an intermediate sequence that is zero for noninteger multiples of Mn, that is

$$x_1(n) = \begin{cases} x(n), & \text{where } n \text{ is a multiple of } M \\ 0, & \text{otherwise.} \end{cases}$$

Therefore, at the decimated sample values $y_D(n) = x(Mn) = x_1(Mn)$. Hence,

$$Y_D(z) = \sum_{n=-\infty}^{\infty} x_1(Mn)z^{-n}.$$

Let $k = Mn$. Then,

$$Y_D(z) = \sum_{k=-\infty}^{\infty} x_1(k)z^{-k/M}.$$

Hence,

$$Y_D(z) = X_1(z^{1/M}).$$

Now, we need to express $X_1(z)$ in terms of $X(z)$. By the definition of $x_1(n)$, we can write

$$x_1(n) = c_M(n)x(n)$$

where, $c_M(n)$ is a sampling function and is defined by

$$c_M(n) = \begin{cases} 1, & \text{whenever } n \text{ is a multiple of } M \\ 0, & \text{otherwise.} \end{cases}$$

So, for example, if $M = 3$ then $c_M(n)$ becomes

$$c_3(n) = 1, 0, 0, 1, 0, 0, 1, 0, 0, 1, \ldots.$$

Mathematically, we say that the sampling function $c_M(n)$ is a sequence of period M, which can be represented by the Fourier series expansion

$$c_M(n) = \frac{1}{M} \sum_{k=0}^{M-1} C(k) \exp\left(\frac{j2\pi kn}{M}\right)$$

where, the Fourier series coefficient $C(k)$ is defined by

$$C(k) = \sum_{n=0}^{M-1} a(n) \exp\left(\frac{-j2\pi kn}{M}\right)$$

on the interval $[0, M)$ with samples at $n = 0, ..., M - 1$. $a(n)$ is defined by

$$a(n) = \begin{cases} 1, & \text{when } n = 0 \\ 0, & \text{otherwise.} \end{cases}$$

So, $C(k) = 1$ for all k. Thus,

$$c_M(n) = \frac{1}{M} \sum_{k=0}^{M-1} \exp\left(\frac{j2\pi kn}{M}\right).$$

Since $x_1(n) = c_M(n)x(n)$, then the z-transform of $x_1(n)$ is given by

$$X_1(z) = \sum_{n=-\infty}^{\infty} x_1(n)z^{-n},$$

or equivalently,

$$X_1(z) = \sum_{n=-\infty}^{\infty} x(n)c_M(n)z^{-n}.$$

Substituting the equation for $c_M(n)$ into the last equation yields

$$X_1(z) = \frac{1}{M} \sum_{k=0}^{M-1} \sum_{n=-\infty}^{\infty} x(n) \left(z \exp\left(\frac{-j2\pi k}{M}\right)\right)^{-n}$$

or equivalently,

$$X_1(z) = \frac{1}{M} \sum_{k=0}^{M-1} X\left(z \exp\left(\frac{-j2\pi k}{M}\right)\right).$$

But $Y_D(z) = X_1\left(z^{1/M}\right)$. Hence,

$$Y_D(z) = \frac{1}{M} \sum_{k=0}^{M-1} X\left(z^{1/M} \exp\left(\frac{-j2\pi k}{M}\right)\right),$$

or equivalently,

$$Y_D(z) = \frac{1}{M} \sum_{k=0}^{M-1} X\left(z^{1/M} W_M^k\right),$$

where $W_M = \exp(\frac{-j2\pi}{M})$ is the (principle) Mth root of unity. The $\frac{1}{M}$th power of z means that the original spectrum is stretched by a factor of M. In the literature, the notation

$$Y_D(z) = X(z)|_{\downarrow M}$$

is occasionally used. If the region of convergence of $Y_D(z)$ includes the points of the form $z = \exp(j\omega)$ where ω is real, then we can replace z with $\exp(j\omega)$ to yield

$$Y_D[\exp(j\omega)] = \frac{1}{M} \sum_{k=0}^{M-1} X\left[\exp(j(\omega - 2\pi k)/M)\right].$$

By suppressing the exponential and, as such, changing the notation so that $Y_D(\omega)$ denotes $Y_D[\exp(j\omega)]$, then we will write

$$Y_D(\omega) = \frac{1}{M} \sum_{k=0}^{M-1} X\left(\frac{\omega - 2\pi k}{M}\right).$$

Hence, a frequency ω_0 in $X(\omega)$ is transformed to M new frequencies $(\omega_0 - 2\pi k)M$, $k = 0, ..., M - 1$, in $Y_D(\omega)$. Now, consider the response $y(n)$ of a decimator to an input sequence $x(n)$, where every other sample is zero. This is simply an example of an intermediate sequence with decimation by 2. So,

$$Y_D(z) = X(z^{1/2}).$$

But using the analysis of the decimator for an arbitrary input $x(n)$, we can say that

$$Y_D(z) = \frac{1}{2}(X(z^{1/2}) + X(-z^{1/2})).$$

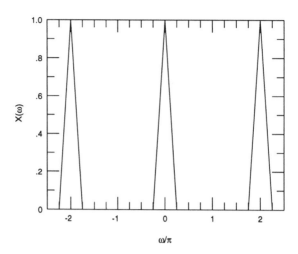

Figure 1.7. Spectrum of input signal.

In the first case, every other input sample was zero valued. Hence, the support of $X(\omega)$ is limited to $-\frac{\pi}{2} \leq \omega < \frac{\pi}{2}$. Under these conditions, decimation by 2 will not create aliasing, that is $X(-z^{1/2}) = X(z^{1/2})$. Therefore, the analysis is consistent for a signal bandlimited to $-\frac{\pi}{2} \leq \omega < \frac{\pi}{2}$. For this example, if the spectrum of the bandlimited input signal is given by Figure 1.7 then the corresponding spectrum of the output of the decimator is given by Figure 1.8. This bandlimited signal could be realized by the use of a lowpass filter with cut-off frequency $\frac{\pi}{M}$ immediately proceeding the decimator as illustrated in Figure 1.9.

This lowpass filter is known as an *anti-aliasing filter*. Recall from Section 1.2, if the sampling interval is increased by a factor of M, then the product of the gains of the decimator and the anti-aliasing filter must equal $\frac{1}{M}$. Since the decimator has a gain of $\frac{1}{M}$, the anti-aliasing filter must have unity gain.

It will be shown later in this chapter that these operators, M-fold decimators and L-fold expanders, are commutative provided that L and M are relatively prime.

1.3.3   Comb filters

Definition 1.3.3.1. *Given the impulse response $h(n)$ of a filter, one can build a comb filter $g(n)$ by inserting $L - 1$ zeros between the coefficients of*

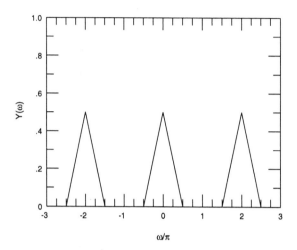

Figure 1.8. Spectrum of decimated signal.

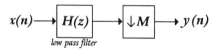

Figure 1.9. Decimator with anti-aliasing filter.

Figure 1.10. Interpolated FIR (IFIR) filter.

$h(n)$, that is

$$g(n) = \begin{cases} h(\frac{n}{L}), & \text{if } n \text{ is a multiple of } L \\ 0, & \text{otherwise.} \end{cases}$$

The impulse response of a comb filter can be represented as

$$g(n) = \sum_{k=-\infty}^{\infty} h(k)\delta_{n,Lk}$$

where, $\delta_{n,Lk}$ is a Kronecker delta function. Taking the z-transform of $g(n)$, we obtain

$$G(z) = \sum_{k=-\infty}^{\infty} \sum_{n=-\infty}^{\infty} h(k)\delta_{n,Lk} \, z^{-n},$$

or equivalently,

$$G(z) = \sum_{k=-\infty}^{\infty} h(k)z^{-Lk},$$

or simply,

$$G(z) = H(z^L).$$

For example, if $H(z) = 1 + 2z^{-1} + 3z^{-2}$ and $L = 3$, then $H(z^3) = 1 + 2z^{-3} + 3z^{-6}$. By inspection, we see that if $H(z)$ is 2π periodic, then $H(z^3)$ is $\frac{2\pi}{3}$ periodic.

An interesting application of comb filters arises in the Interpolated FIR (IFIR) design technique. The idea is quite simple. Assume some filter $G(z)$ meets the desired specifications. Is there a way to use comb filters to reduce the number of computations required to realize the filter? Consider Figure 1.10, where $I(z)$ is a lowpass filter that eliminates the $L - 1$ unwanted passbands generated from $G(z^L)$.

If the filter $G(z)$ meets the desired specifications, the cascade of $I(z)$ and $G(z^L)$ will also meet the specifications. If L is a small number, then this implementation works quite well. But if L is too large, then the transition band of $I(z)$ becomes quite narrow and $I(z)$ will dominate the computational complexity, that is, $I(z)$ will dominate the number of multiplications

and additions required. The following example demonstrates the lower costs achievable with an IFIR filter implementation.

Example 1.3.3.1. Suppose we wish to design a linear phase FIR filter with specifications:

$$\text{peak passband ripple: } \delta_1 = 0.001$$

$$\text{peak stopband ripple: } \delta_2 = 0.001$$

$$\text{passband edge: } \omega_p = 0.01\pi$$

$$\text{stopband edge: } \omega_s = 0.02\pi.$$

Method 1: Equiripple Design. The normalized transition bandwidth, Δf, is given by:

$$\Delta f = \frac{\omega_s - \omega_p}{2\pi} = \frac{0.01\pi}{2\pi} = 0.005.$$

Then, we can estimate the filter order by:

$$N = \left\lceil \frac{-13 - 20 \log_{10} (\delta_1 \delta_2)^{0.5}}{14.6 \, \Delta f} \right\rceil.$$

So,

$$N = \left\lceil \frac{-13 - 20 \log_{10} (10^{-6})^{0.5}}{14.6 \, (0.005)} \right\rceil = 644.$$

The number of multiplications required is given by

$$\# \text{ multiplies} = \left\lceil \frac{N+1}{2} \right\rceil = \left\lceil \frac{645}{2} \right\rceil = 323,$$

and the number of additions required is given by

$$\# \text{ adds} = N = 644.$$

Method 2: IFIR Design with a stretching factor of 15. Let us first consider the filter $G(z^{15})$. Its specifications become

$$\text{peak passband ripple: } \delta_{1_G} = \frac{\delta_1}{2} = 0.0005$$

$$\text{peak stopband ripple: } \delta_{2_G} = \delta_2 = 0.001$$

$$\text{passband edge: } \omega_{p_G} = (0.01\pi)15 = 0.15\pi$$

$$\text{stopband edge: } \omega_{s_G} = (0.02\pi)15 = 0.3\pi.$$

Again consider an equiripple design. The normalized transition bandwidth, $(\Delta f)_G$, is given by:

$$(\Delta f)_G = \frac{\omega_{s_G} - \omega_{p_G}}{2\pi} = \frac{0.15\pi}{2\pi} = 0.075.$$

Then, we can estimate the filter order by:

$$N_G = \left\lceil \frac{-13 - 20 \log_{10}\ \left(\delta_{1_G}\delta_{2_G}\right)^{0.5}}{14.6\ (\Delta f)_G} \right\rceil .$$

So,

$$N_G = \left\lceil \frac{-13 - 20 \log_{10}\ \left(5 \times 10^{-7}\right)^{0.5}}{14.6\ (0.075)} \right\rceil = 46.$$

The number of multiplications required for $G(z)$ is given by

$$(\# \text{ multiplies})_G = \left\lceil \frac{N_G + 1}{2} \right\rceil = \left\lceil \frac{47}{2} \right\rceil = 24,$$

and the number of additions required is given by

$$(\# \text{ adds})_G = N_G = 46.$$

Let us now consider the filter $I(z)$. Then, its specifications become

peak passband ripple: $\delta_{1_I} = \frac{\delta_1}{2} = 0.0005$

peak stopband ripple: $\delta_{2_I} = \delta_2 = 0.001$

passband edge: $\omega_{p_I} = 0.01\pi$

stopband edge: $\omega_{s_I} = \frac{2\pi}{15} - 0.02\pi = 0.113\pi.$

Considering an equiripple design, the normalized transition bandwidth, $(\Delta f)_I$, is given by:

$$(\Delta f)_I = \frac{\omega_{s_I} - \omega_{p_I}}{2\pi} = \frac{0.103\pi}{2\pi} = 0.0515.$$

Then, we can estimate the filter order for $I(z)$ by:

$$N_I = \left\lceil \frac{-13 - 20 \log_{10}\ \left(\delta_{1_I}\delta_{2_I}\right)^{0.5}}{14.6\ (\Delta f)_I} \right\rceil .$$

So,

$$N_I = \left\lceil \frac{-13 - 20 \log_{10}\ \left(5 \times 10^{-7}\right)^{0.5}}{14.6\ (0.0515)} \right\rceil = 67.$$

The number of multiplications required for $I(z)$ is given by

$$(\# \text{ multiplies})_I = \left\lceil \frac{N_I + 1}{2} \right\rceil = \left\lceil \frac{68}{2} \right\rceil = 34,$$

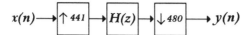

Figure 1.11. Conversion from studio work to CD mastering.

and the number of additions required is given by

$$(\# \text{ adds})_I = N_I = 67.$$

Hence,

$$(\# \text{ multiplies})_{IFIR} = (\# \text{ multiplies})_G + (\# \text{ multiplies})_I = 113$$
$$(\# \text{ adds})_{IFIR} = (\# \text{ adds})_G + (\# \text{ adds})_I = 58.$$

So, using the IFIR design technique, we can significantly decrease the computational complexity over conventional equiripple filter design.

1.4 Interchanging basic building blocks

This section first presents the conditions for interchanging decimators and expanders. Secondly, we consider the noble identities, an approach for interchanging filters with the basic building blocks for decimation and expansion.

1.4.1 Interchanging decimators and expanders

Sampling rate conversion is important for many signal processing applications. For example, conversion between the three standards for digital audio, which are 48KHz for studio work, 44.1KHz for CD mastering (both digital tape and compact discs), and 32 KHz for broadcasting digital audio. To convert from the studio work frequency to the CD mastering frequency, one could use the arrangement in Figure 1.11. Thus, sampling rate conversion will require the cascade of an L-fold expander and an M-fold decimator, separated by a filter. Thus, by cascading an L-fold expander and an M-fold decimator, one obtains a rational sampling rate change with rate L/M.

Under what conditions can the decimator and the expander be interchanged? First consider the configuration in which the expander proceeds the decimator as depicted in Figure 1.12.

Writing the associated elemental equations yields

$$T_1(z) = X\left(z^L\right)$$

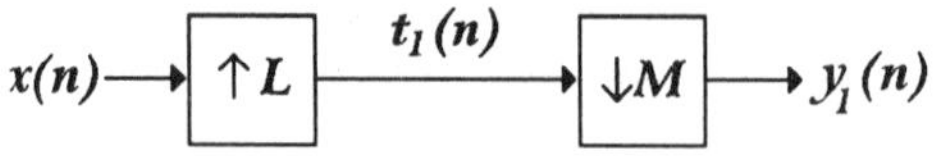

$$x(n) \longrightarrow \boxed{\uparrow L} \xrightarrow{\quad t_l(n) \quad} \boxed{\downarrow M} \longrightarrow y_l(n)$$

Figure 1.12. L-fold expander proceeding M-fold decimator.

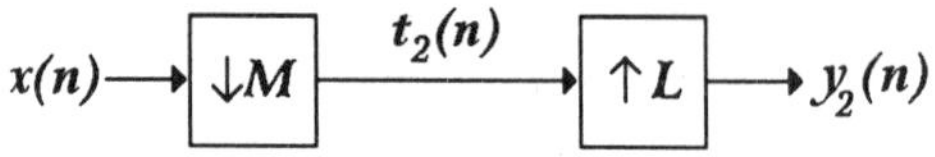

$$x(n) \longrightarrow \boxed{\downarrow M} \xrightarrow{\quad t_2(n) \quad} \boxed{\uparrow L} \longrightarrow y_2(n)$$

Figure 1.13. M-fold decimator proceeding L-fold expander.

and

$$Y_1(z) = \frac{1}{M} \sum_{k=0}^{M-1} T_1 \left[z^{1/M} \exp\left(\frac{-j2\pi k}{M} \right) \right].$$

Hence,

$$Y_1(z) = \frac{1}{M} \sum_{k=0}^{M-1} X \left(\left[z^{1/M} \exp\left(\frac{-j2\pi k}{M} \right) \right]^L \right),$$

or equivalently,

$$Y_1(z) = \frac{1}{M} \sum_{k=0}^{M-1} X \left[z^{L/M} \exp\left(\frac{-j2\pi kL}{M} \right) \right].$$

Second, consider the configuration in which the decimator proceeds the expander as depicted in Figure 1.13. Writing the associated elemental equations yields

$$T_2(z) = \frac{1}{M} \sum_{k=0}^{M-1} X \left[z^{1/M} \exp\left(\frac{-j2\pi k}{M} \right) \right]$$

and

$$Y_2(z) = T_2 \left(z^L \right).$$

Hence,

$$Y_2(z) = \frac{1}{M} \sum_{k=0}^{M-1} X \left[z^{L/M} \exp\left(\frac{-j2\pi k}{M} \right) \right].$$

Therefore, an L-fold expander can be interchanged with an M-fold decimator if and only if $Y_1(z) = Y_2(z)$. By inspection, this will only occur when $\exp(-\frac{j2\pi kL}{M}), k = 0, ..., M - 1$, generates the Mth roots of unity. The following theorem provides the conditions on L and M for this to occur.

Theorem 1.4.1.1. *The set* $\{\exp(-\frac{j2\pi kL}{M})|k = 0, ..., M - 1\}$ *equals the set of the Mth roots of unity, if and only if L and M are relatively prime, that is*

$$\gcd(L, M) = 1.$$

Proof: Case 1: Assume $\exp(-\frac{j2\pi kL}{M}), k = 0, ..., M - 1$, generates all the Mth roots of unity. Assume, for contradiction purposes, that $\gcd(L, M) = d > 1$. Then, there exist relatively prime integers L_1 and M_1, such that $L = dL_1$ and $M = dM_1$. Hence, $\exp(-\frac{j2\pi kL}{M}) = \exp(-\frac{j2\pi kL_1}{M_1})$. But, $\exp(-\frac{j2\pi kL_1}{M_1})$ will generate at most M_1 roots of unity. Since $M_1 < M$, a contradiction occurs.

Case 2: Assume L and M are relatively prime. Assume, for contradiction purposes, that $\exp(-\frac{j2\pi kL}{M}), k = 0, ..., M - 1$, generates $M_1 < M$ roots of unity, where $M = \frac{e_1}{f_1}M_1$ for some relatively prime positive integers e_1 and f_1. Then, $\exp(-\frac{j2\pi kL}{M}) = \exp(-\frac{j2\pi}{M_1}(\frac{f_1 kL}{e_1}))$. Since k can take on all values from 0 to $M - 1$ and since $\gcd(e_1, f_1) = 1$, then $e_1 | L$. Since $e_1 | L$ and $e_1 | M$, then $\gcd(L, M) = e_1$. Therefore, a contradiction occurs. ∎

Hence, an L-fold expander and an M-fold decimator commute provided L and M are relatively prime. Let us illustrate this result with a couple examples. Let $L = 2$ and $M = 3$, then $\gcd(2, 3) = 1$. Thus, $\exp(-\frac{j2\pi kL}{M})$ becomes $\exp(-\frac{j4\pi k}{3})$. So, for $k = 0, 1, 2$ we have the following terms

$$\exp(-j0) = 1; \quad \exp\left(-\frac{j4\pi}{3}\right); \quad \exp\left(-\frac{j8\pi}{3}\right) = \exp\left(-\frac{j2\pi}{3}\right)$$

while, $\exp(-\frac{j2\pi k}{M})$ becomes $\exp(-\frac{j2\pi k}{3})$. So, for $k = 0, 1, 2$ we have the following terms

$$\exp(-j0) = 1; \exp\left(-\frac{j2\pi}{3}\right); \exp\left(-\frac{j4\pi}{3}\right).$$

Therefore, as expected, the two expressions are equivalent.

Now, as a counterexample suppose $L = 2$ and $M = 4$, then $\gcd(2, 4) = 2$. Thus, $\exp(-\frac{j2\pi kL}{M})$ becomes $\exp(-j\pi k)$. So, for $k = 0, 1, 2, 3$ we have the following terms

$$\exp(-j0) = 1; \exp(-j\pi) = -1; \exp(-j2\pi) = 1; \exp(-j3\pi) = -1,$$

while $\exp(-\frac{j2\pi k}{M})$ becomes $\exp(-\frac{j\pi k}{2})$. So, for $k = 0, 1, 2, 3$ we have the following terms

$$\exp(-j0) = 1; \exp\left(-\frac{j\pi}{2}\right) = -j; \exp(-j\pi) = -1; \exp\left(-\frac{j3\pi}{2}\right) = j.$$

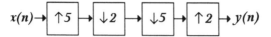

Figure 1.14. A more complicated example.

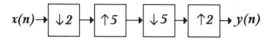

Figure 1.15. Equivalent version of Figure 1.14.

Thus, as expected, these two expressions are not equivalent.

Now, let us consider a slightly more complicated example, which is depicted in Figure 1.14. Find an expression for $y(n)$ in terms of $x(n)$.

Applying Theorem 1.4.1.1, we can interchange the first two blocks, since the expansion ratio of 5 is relatively prime to the decimation ratio of 2. This will yield Figure 1.15. Now, consider the two middle blocks as depicted in Figure 1.16. Then,

$$T(z) = \frac{1}{5} \sum_{k=0}^{4} S\left(z \exp\left[-j\frac{2\pi(5k)}{5}\right]\right)$$

or equivalently,

$$T(z) = \frac{1}{5} \sum_{k=0}^{4} S(z)$$

or simply,

$$T(z) = S(z).$$

So, we can eliminate the two middle blocks without effecting the output. Hence, Figure 1.17 shows the simplified version of Figure 1.14.

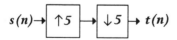

Figure 1.16. Middle blocks.

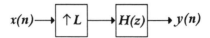

Figure 1.17. Simplified version of Figure 1.14.

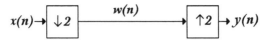

Figure 1.18. Expander followed by interpolation filter.

By inspection,

$$w(n) = x(2n)$$

and

$$y(n) = \begin{cases} w(\tfrac{n}{2}) & \text{for} \quad n \text{ even} \\ 0 & \text{for} \quad n \text{ odd.} \end{cases}$$

Therefore,

$$y(n) = \begin{cases} x(n) & \text{for} \quad n \text{ even} \\ 0 & \text{for} \quad n \text{ odd.} \end{cases}$$

1.4.2 Noble identities

In most applications involving interpolation, an interpolation filter follows an expander as in Figure 1.18.

The filter must *handle* data at an increased sampling rate of L. We would like to find ways to reduce the speed at which the filter operates. One way to do this is to interchange the expander and the filter in an equivalent system as in Figure 1.19.

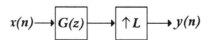

Figure 1.19. Filter proceeding expander.

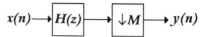

Figure 1.20. Decimation filter followed by decimator.

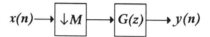

Figure 1.21. Decimator proceeding filter.

Let us discuss the conditions under which they are equivalent. Similarly, in many applications involving decimators, a decimation filter precedes the decimator as in Figure 1.20.

Analogously, we would like to allow $H(z)$ to operate at the *slower rate* of the decimator. Is there any equivalent system to allow this interchange? What must $G(z)$ be for the system depicted in Figure 1.21?

If $H(z)$ is a comb filter, then can the building blocks be interchanged?

1.4.2.1 Interchanging filters and expanders

Consider the configuration depicted in Figure 1.22. Writing the elemental equations, we obtain

$$V(z) = X(z^L)$$

and

$$Y(z) = G(z^L)V(z).$$

Hence,

$$Y(z) = G(z^L)X(z^L).$$

But, this equation could also be interpreted as Figure 1.23. These two equivalent block diagrams in Figure 1.24 present a systematic approach for

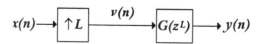

Figure 1.22. Expander followed by comb filter.

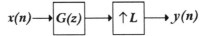

Figure 1.23. Filter proceeding L-fold expander.

Figure 1.24. Noble identity for expanders.

interchanging filters with expanders and together they will be referred to as the noble identity for expanders. Please note that the noble identity for expanders does not, in general, apply if $G(z)$ is irrational.

1.4.2.2 Interchanging filters and decimators

Consider the configuration depicted in Figure 1.25. Writing the elemental equations,

$$V(z) = G(z^M)X(z)$$

and

$$Y(z) = \frac{1}{M} \sum_{k=0}^{M-1} V\left[z^{1/M} \exp\left(\frac{-j2\pi k}{M}\right)\right].$$

Hence,

$$Y(z) = \frac{1}{M} \sum_{k=0}^{M-1} G\left(\left[z^{1/M}\exp\left(\frac{-j2\pi k}{M}\right)\right]^M\right) X\left(z^{1/M}\exp\left(\frac{-j2\pi k}{M}\right)\right),$$

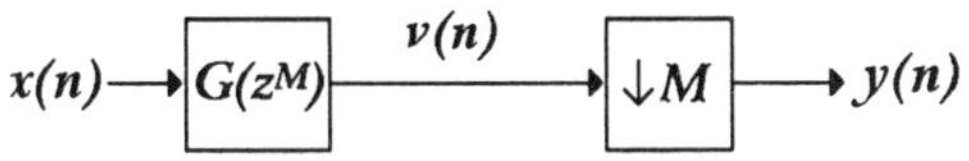

$$x(n) \longrightarrow \boxed{G(z^M)} \xrightarrow{\;v(n)\;} \boxed{\downarrow M} \longrightarrow y(n)$$

Figure 1.25. Comb filter proceeding M-fold decimator.

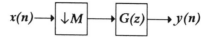

Figure 1.26. Decimator proceeding filter.

Figure 1.27. Noble identity for decimators.

or equivalently,

$$Y(z) = \frac{G(z)}{M} \sum_{k=0}^{M-1} X\left[z^{1/M}\exp\left(\frac{-j2\pi k}{M}\right)\right].$$

But this equation could be interpreted as Figure 1.26. These two equivalent block diagrams, depicted in Figure 1.27, present a systematic approach for interchanging filters with decimators and together they will be referred to as the noble identity for decimators. Please note that the noble identity for decimators does not, in general, apply if $G(z)$ is irrational.

1.5   A filter bank example

Consider the following example depicted in Figure 1.28 to gain some intuition into the importance of polyphase and decimators and expanders for multirate filter banks.

As part of this example, let us examine the inputs, outputs, and intermediate results for this simple filter bank. For this example, let us assume that the first nonzero input sample to the filter bank will be designated $x(0)$. Then,

$x(n):$	$x(0)$	$x(1)$	$x(2)$	$x(3)$	$x(4)$	$x(5)$	$x(6)$	...
$x_0(n):$	$x(0)$		$x(2)$		$x(4)$		$x(6)$	...
$x_1(n):$	0		$x(1)$		$x(3)$		$x(5)$	...
$y_0(n):$	$x(0)$	0	$x(2)$	0	$x(4)$	0	$x(6)$	...
$y_1(n):$	0	0	$x(1)$	0	$x(3)$	0	$x(5)$	...
$w(n):$	0	$x(0)$	$x(1)$	$x(2)$	$x(3)$	$x(4)$	$x(5)$	... .

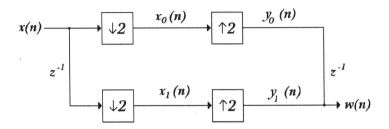

Figure 1.28. A simple filter bank.

Thus, the upper decimator passes only even-numbered samples, while the lower decimator passes only odd-numbered samples. Therefore, the outputs of the decimators correspond to the polyphase components of the filter bank.

1.6 Problems

1. Let an input signal $x(n)$ be periodic with period N, that is, $x(n) = x(n + N)$. Let $y(n)$ be an M-fold decimated version of $x(n)$, that is, $y(n) = x(Mn)$. Show that $y(n)$ is periodic with some period P, that is, $y(n) = y(n + P)$. What is the smallest value of P in terms of M and N?

2. In the systems depicted in Figures 1.12 and 1.13, derive the sequence domain equations for their outputs in terms of their input, that is, $y_1(n)$ and $y_2(n)$ in terms of $x(n)$. Then, prove that these two expressions yield the same result if and only if L and M are relatively prime.

3. Prove the following:

$$((A(z))_{\uparrow K} B(z))_{\downarrow KL} = (A(z)(B(z))_{\downarrow K})_{\downarrow L}.$$

4. Let the input $x(n)$ be defined by $x(n) = \delta_{0,n}$. Apply this input to a decimator and show that the decimator could be interpreted as a periodic time-invariant system.

5. Some authors have proposed a Type-III polyphase representation using the following definition: Given a sequence $x(n)$, then its Type-III polyphase components are defined as

$$x_k(n) = x(Mn - k).$$

How are the Type-I and Type-II polyphase components related to Type-III polyphase components?

6. In example 1.3.3.1, we compared the complexity of a conventional equiripple filter design with an IFIR design with a stretching factor of 15. Repeat the analysis for a stretching factor of 40.

Chapter 2

Multidimensional Multirate Signal Processing

2.1 Introduction

This chapter extends the basic concepts of multirate signal processing to multidimensional multirate signal processing. Examples of multidimensional signals include images in two dimensions and video in three dimensions. Good reference texts for background material on multidimensional signal processing are Dudgeon and Mersereau[16] and Lim[26]. The important concept of sampling is related to the mathematics of lattices; see for example Cassels[4]. The engineering analysis of sampling began in the classic paper by Petersen and Middleton[38] and later extended to include decimation and expansion considerations by Mersereau and Speake[31] and Dubois[15]. The multidimensional z-transform is carefully described by Viscito and Allebach[55]. The idea of the Smith form was first articulated by Smith[43]. Many recent papers have dealt with applications of the Smith form to the multidimensional DFT (Guessoum and Mersereau[20]), to the multidimensional DCT (Gündüzhan *et al.*[21]), and for the development of multirate CAD tools (Evans[18]). Some of the concepts developed in this chapter are also discussed in the text by Vaidyanathan[49].

Section 2.2 presents a framework for the study of multidimensional multirate signal processing and it introduces two important representations for multidimensional signals. Section 2.3 presents the multidimensional building blocks. Section 2.4 provides ways to interchange the multidimensional building blocks.

2.2 Multidimensional framework

A multidimensional signal is a signal of more than one variable. This section systematically presents concepts that act as a framework for our study

29

of the application of multirate systems to multidimensional signals. These concepts include sampling of a multidimensional signal, which involves the mathematical concept of a sampling lattice, and introduce multidimensional sampled signals by way of multidimensional z-transforms and multidimensional Fourier transforms. Now, we will discuss the mathematics needed to describe the sampling considerations of multidimensional multirate.

2.2.1 Sampling lattices

Sampling a multidimensional signal is more complicated than a one-dimensional signal because of the many ways to choose the sampling geometry. Sampling points could be arranged on a rectangular grid in a straightforward manner, but many times there are more efficient ways to sample multidimensional signals.

Some applications are more suitable for nonrectangular sampling than rectangular sampling. For example, the testing of a phased-array antenna requires the measurement of the electric field on a plane in the near field of the antenna. Since phased-array antennas are typically designed with an hexagonal arrangement of elements, the resulting processing is more accurate if hexagonal sampling is used for the measurements.

One reason to consider nonrectangular sampling is to minimize the number of points needed to characterize an M-dimensional hypervolume. Moreover, if the functions of interest are bandlimited over a circular region, then Figure 2.1 shows that significant savings are possible if hexagonal sampling is used instead of rectangular sampling.

In order to precisely describe both rectangular and nonrectangular sampling, we need a convenient way to describe an arbitrary sampling geometry. To do this activity, we must appeal to the language of linear algebra in order to present the mathematical theory of sampling lattices.

2.2.1.1 Linear independence and sampling lattices

The idea of a vector in an integer-valued k-dimensional space is a generalization of vectors in a plane by integer-valued Cartesian coordinates. This leads to the following definition.

Definition 2.2.1.1. *The set of all k-dimensional integer vectors will be called the* fundamental lattice *and it will be denoted $\mathcal{N}$. That is,*

$$\mathcal{N} = \{\mathbf{r} = [r_1, r_2, \ldots, r_k]^T \mid r_i \text{ is an integer}\}.$$

The set of all k-dimensional real vectors will be denoted $\mathcal{E}$.

Now, let us review a few definitions related to sums of vectors.

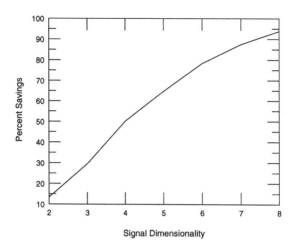

Figure 2.1. Percent savings: hexagonal versus rectangular sampling.

Definition 2.2.1.2. *A linear combination of* N *vectors* $\{\mathbf{r}_1, \ldots \mathbf{r}_N\} \in \mathcal{N}$ *is an expression of the form*

$$\sum_{i=1}^{N} n_i \mathbf{r}_i$$

where n_i, $i = 1, \ldots, N$, *are integers and are called coefficients. The set of vectors* $\{\mathbf{r}_1, \ldots \mathbf{r}_N\}$ *is said to be linearly independent if*

$$\sum_{i=1}^{N} n_i \mathbf{r}_i = \mathbf{0} \Rightarrow n_i = 0 \text{ for all } i.$$

If the set of vectors $\{\mathbf{r}_1, \ldots \mathbf{r}_N\}$, *is linearly independent, then the totality of vectors of the form*

$$\left\{ \sum_{i=1}^{N} n_i \mathbf{r}_i \middle| n_i \text{ is an integer} \right\}$$

is called a N-dimensional lattice and it will be denoted by $\mathcal{R}$. *In other words, the space* $\mathcal{R}$ *spanned by the set of vectors* $\{\mathbf{r}_1, \ldots \mathbf{r}_N\}$ *is the space consisting of all linear combinations of the vectors, that is,*

$$\text{span}(\mathcal{R}) = \left\{ \sum_{i=1}^{N} n_i \mathbf{r}_i \middle| n_i \text{ is an integer} \right\}.$$

The set of vectors $\{\mathbf{r}_1, \ldots \mathbf{r}_N\}$ *is a basis for* $\mathcal{R}$ *if they are linearly indepen-dent and the space is spanned by* $\{\mathbf{r}_1, \ldots \mathbf{r}_N\}$ *is equal to* $\mathcal{R}$. *Then, we say that* $\mathcal{R}$ *has dimension* N.

To better understand the definition of the lattice, consider the following illustration in two dimensions. Let

$$\mathbf{r}_1 = \begin{bmatrix} r_{11} \\ r_{21} \end{bmatrix}, \ \mathbf{r}_2 = \begin{bmatrix} r_{12} \\ r_{22} \end{bmatrix}, \text{ and } \mathbf{n} = \begin{bmatrix} n_1 \\ n_2 \end{bmatrix}.$$

Then,

$$n_1 \mathbf{r}_1 + n_2 \mathbf{r}_2 = n_1 \begin{bmatrix} r_{11} \\ r_{21} \end{bmatrix} + n_2 \begin{bmatrix} r_{12} \\ r_{22} \end{bmatrix},$$

or equivalently,

$$n_1 \mathbf{r}_1 + n_2 \mathbf{r}_2 = \begin{bmatrix} n_1 r_{11} + n_2 r_{12} \\ n_1 r_{21} + n_2 r_{22} \end{bmatrix}.$$

This can be rewritten as a matrix-vector product, *i.e.*

$$n_1 \mathbf{r}_1 + n_2 \mathbf{r}_2 = \begin{bmatrix} r_{11} & r_{12} \\ r_{21} & r_{22} \end{bmatrix} \begin{bmatrix} n_1 \\ n_2 \end{bmatrix} = \begin{bmatrix} \mathbf{r}_1 & \mathbf{r}_2 \end{bmatrix} \mathbf{n} = \mathbf{R}\mathbf{n}$$

where the matrix $\mathbf{R}$ is called the *sampling matrix*. In general, let $\mathbf{r}_i$ be the ith column of the matrix $\mathbf{R}$, that is

$$\mathbf{R} = [\mathbf{r}_1, \mathbf{r}_2, \ldots, \mathbf{r}_N],$$

then the sampling matrix $\mathbf{R}$ is said to generate the lattice $\mathcal{R}$. As such, the lattice $\mathcal{R}$ is also given by

$$\mathcal{R} = \text{LAT}(\mathbf{R}) = \{\mathbf{m} \in \mathcal{N} | \ \mathbf{m} = \mathbf{R}\mathbf{n} \text{ for } \mathbf{n} \in \mathcal{N} \ \}.$$

If $\mathbf{R}$ is the identity matrix, then each $\mathbf{r}_i$ is a unit vector pointing in the ith direction, and the resulting lattice, $\mathcal{R}$, is the *fundamental lattice* $\mathcal{N}$.

Let us present some examples of sampling lattices using black dots to represent the lattice points and white circles to represent points in $\mathcal{N}$ that are not in $\text{LAT}(\mathbf{R})$.

Example 2.2.1.1. Consider rectangular sampling defined by the sampling matrix $\mathbf{R} = \begin{bmatrix} 2 & 0 \\ 0 & 3 \end{bmatrix}$. It is depicted in Figure 2.2.

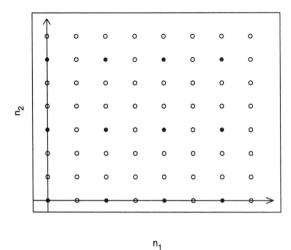

Figure 2.2. Lattice structure using rectangular sampling matrix.

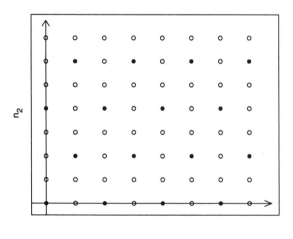

Figure 2.3. Lattice structure using hexagonal sampling matrix.

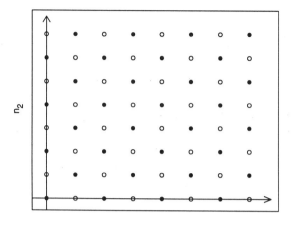

Figure 2.4. Lattice structure with quincunx sampling matrix.

Example 2.2.1.2. Consider hexagonal sampling defined by the sampling matrix $\mathbf{R} = \begin{bmatrix} 1 & 1 \\ 2 & -2 \end{bmatrix}$. It is depicted in Figure 2.3.

Example 2.2.1.3. Consider quincunx sampling defined by the sampling matrix $\mathbf{R} = \begin{bmatrix} 1 & 1 \\ -1 & 1 \end{bmatrix}$. It is depicted in Figure 2.4.

For a given sampling matrix $\mathbf{R}$, the corresponding Fourier domain sampling matrix is $2\pi\mathbf{R}^{-T}$ and the lattice it generates is called the reciprocal lattice.

2.2.1.2   Nonuniqueness and unimodular matrices

The matrix that generates a lattice is not unique. As we will see later in this subsection, the following matrices generate the same lattice.

$$\mathbf{R} = \begin{bmatrix} 1 & 0 \\ 2 & -4 \end{bmatrix} \text{ and } \mathbf{S} = \begin{bmatrix} 1 & 1 \\ 2 & -2 \end{bmatrix}.$$

The theory underlying the nonuniqueness of these sampling lattices is based on unimodular matrices. Thus, in order to discuss the nonuniqueness of sampling lattices, we must first briefly discuss unimodular matrices.

Definition 2.2.1.3. *An integer-valued matrix* **A** *is a unimodular matrix if* $|det\ \mathbf{A}| = 1$.

The following theorems provide properties of these matrices.

Theorem 2.2.1.1. *If* **A** *is an integer-valued unimodular matrix, then* $\mathbf{A}^{-1}$ *exists and is an integer-valued unimodular matrix.*

Proof: Let **A** be an integer-valued unimodular matrix. Moreover, since $\det \mathbf{A} \neq 0$, then $\mathbf{A}^{-1}$ exists and $\mathbf{A}\mathbf{A}^{-1} = \mathbf{I}$. Since $\det \mathbf{A} \det \mathbf{A}^{-1} = 1$, then $|\det \mathbf{A}||\det \mathbf{A}^{-1}| = 1$. Since **A** is unimodular, then $|\det \mathbf{A}| = 1$. Hence, $|\det \mathbf{A}^{-1}| = 1$. By definition,

$$\mathbf{A}^{-1} = \frac{\text{adjugate}\,(\mathbf{A})}{\det \mathbf{A}},$$

where the (i, j) element of the adjugate($\mathbf{A}$) is equal to the cofactor of the (j, i) element of ($\mathbf{A}$). Since **A** is unimodular, then

$$\mathbf{A}^{-1} = \pm\ \text{adjugate}\,(\mathbf{A}).$$

Since **A** is an integer-valued matrix, then adjugate($\mathbf{A}$) is an integer-valued matrix. Therefore, $\mathbf{A}^{-1}$ is an integer-valued unimodular matrix. ∎

With this background on integer-valued unimodular matrices, we are ready to discuss the issue of nonuniqueness of sampling lattices.

Theorem 2.2.1.2. *Given a nonsingular matrix* **M** *and an integer-valued unimodular matrix* **V**, *then LAT(MV) = LAT(M).*

Proof: If $\mathbf{x} \in \text{LAT}(\mathbf{M})$, then there exists an $\mathbf{n} \in \mathcal{N}$, such that $\mathbf{x} = \mathbf{Mn} = \mathbf{MVV}^{-1}\mathbf{n} = \mathbf{MVm}$, where $\mathbf{m} = \mathbf{V}^{-1}\mathbf{n}$. Since **V** is an integer-valued unimodular matrix, then $\mathbf{V}^{-1}$ is also an integer-valued unimodular matrix by Theorem 2.2.1.1. Since $\mathbf{V}^{-1}$ is integer-valued, then $\mathbf{m} = \mathbf{V}^{-1}\mathbf{n}$ is integer-valued. Therefore, $\mathbf{x} \in \text{LAT}(\mathbf{MV})$. Therefore, $\text{LAT}(\mathbf{M}) \subseteq \text{LAT}(\mathbf{MV})$. Obviously, $\text{LAT}(\mathbf{MV}) \subset \text{LAT}(\mathbf{M})$. Since **M** and **MV** generate the same lattice, then $\text{LAT}(\mathbf{MV}) = \text{LAT}(\mathbf{M})$. ∎

For notation purposes, let $J(\mathbf{M}) = |\det \mathbf{M}|$, the absolute value of the determinant of the sampling matrix **M**.

Theorem 2.2.1.3. *Given a nonsingular matrix* **M** *and an integer-valued unimodular matrix* **V**. *Then,* $J(\mathbf{M}) = J(\mathbf{MV}) = J(\mathbf{VM})$.

Proof: Since $J(\mathbf{M}) = |\det \mathbf{M}|$, then

$$J(\mathbf{MV}) = J(\mathbf{M})J(\mathbf{V}) = J(\mathbf{V})J(\mathbf{M}) = J(\mathbf{VM}).$$

Since $\mathbf{V}$ is an integer-valued unimodular matrix, then $J(\mathbf{V}) = 1$. Hence,

$$J(\mathbf{M}) = J(\mathbf{MV}) = J(\mathbf{VM}).$$

$\blacksquare$

     Therefore, $J(\mathbf{M})$ is unique and independent of the choice of basis vectors. Moreover, $J(\mathbf{M})$ can be interpreted geometrically as the k-dimensional volume of the parallelepiped defined by $\mathbf{M}$. Sometimes $J(\mathbf{M})$ is called the sampling density. Now, consider the following sampling matrices

$$\mathbf{R} = \begin{bmatrix} 1 & 0 \\ 2 & -4 \end{bmatrix} \text{ and } \mathbf{S} = \begin{bmatrix} 1 & 1 \\ 2 & -2 \end{bmatrix}$$

where $J(\mathbf{R}) = J(\mathbf{S}) = 4$. Since $\mathbf{RE} = \mathbf{S}$ for unimodular matrix

$$\mathbf{E} = \begin{bmatrix} 1 & 1 \\ 0 & 1 \end{bmatrix},$$

then $\mathbf{R}$ and $\mathbf{S}$ generate the same sampling lattice. In this case, the sampling lattice is known as an hexagonal sampling lattice.

2.2.1.3   Unit cells and fundamental parallelepipeds

Definition 2.2.1.4. *Given an integer-valued matrix* $\mathbf{R}$, *a unit cell includes one lattice point from LAT($\mathbf{R}$) and J($\mathbf{R}$)-1 adjacent points in $\mathcal{N}$ that are not in LAT($\mathbf{R}$).*

     If these unit cells are periodically replicated on LAT($\mathbf{R}$), then the entire space is tiled with no overlap. Thus, the unit cell is a footprint that characterizes the sampling lattice.

Definition 2.2.1.5. *Given an integer-valued matrix* $\mathbf{R}$, *the fundamental parallelepiped of lattice LAT($\mathbf{R}$), denoted FPD($\mathbf{R}$), is the unit cell that includes the origin and is bounded by all lattice points one positive unit away. Formally, the fundamental parallelepiped is given by*

$$FPD(\mathbf{R}) = \{\mathbf{y} \in \mathcal{E} \mid \mathbf{y} = \mathbf{Rx} \text{ for all } \mathbf{x} \in [0,1)^k\},$$

where $\mathcal{E}$ is the set of all k-dimensional real vectors.

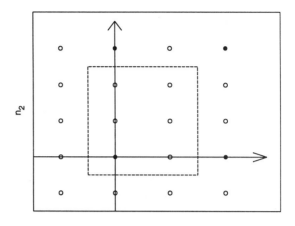

Figure 2.5. Fundamental parallelepiped example.

Consider the following example as an illustration of the concept of fundamental parallelepipeds. Assume the sampling matrix $\mathbf{R} = \begin{bmatrix} 2 & 0 \\ 0 & 3 \end{bmatrix}$. Then, fundamental parallelepiped is given in Figure 2.5. By inspection, there are $J(\mathbf{R}) = 6$ points in the fundamental parallelepiped.

Theorem 2.2.1.4. *Given an integer-valued unimodular matrix* $\mathbf{U}$ *and an integer-valued diagonal matrix* Λ, *then*

$$FPD(\mathbf{U}\Lambda) = \mathbf{U}\ FPD(\Lambda).$$

Proof: By the definition of the fundamental parallelepiped,

$$\text{FPD}(\mathbf{U}\Lambda) = \{\mathbf{y}\ \in \mathcal{E}\ |\ \mathbf{y} = \mathbf{U}\Lambda\mathbf{x} \text{ for all } \mathbf{x} \in [0,1)^k\}.$$

Let $\mathbf{z} = \Lambda\mathbf{x}$. Then, $\mathbf{x} = \Lambda^{-1}\mathbf{z}$. Hence,

$$\text{FPD}(\mathbf{U}\Lambda) = \{\mathbf{y}\ \in \mathcal{E}\ |\ \mathbf{y} = \mathbf{U}\mathbf{z} \text{ for all } \Lambda^{-1}\mathbf{z} \in [0,1)^k\}.$$

Since $\mathbf{U}$ is an integer-valued unimodular matrix, then $\mathbf{U}\mathbf{z}$ is an integer-valued vector if and only if $\mathbf{z}$ is an integer vector. Therefore,

$$\text{FPD}(\mathbf{U}\Lambda) = \mathbf{U}\{\mathbf{y}\ \in \mathcal{E}\ |\ \mathbf{y} = \mathbf{z} \text{ for all } \Lambda^{-1}\mathbf{z} \in [0,1)^k\}.$$

But $\mathbf{z} = \Lambda\mathbf{x}$. Hence,

$$\text{FPD}(\mathbf{U}\Lambda) = \mathbf{U}\{\mathbf{y} \in \mathcal{E} \mid \mathbf{y} = \Lambda\mathbf{x} \text{ for all } \mathbf{x} \in [0,1)^k\}$$

or equivalently,

$$\text{FPD}(\mathbf{U}\Lambda) = \mathbf{U}\,\text{FPD}(\Lambda).$$

∎

Sometimes in the literature, authors will refer to the symmetric parallelepiped, denoted SPD($\mathbf{R}$). It is defined by

$$\text{SPD}(\mathbf{R}) = \{\mathbf{y} \in \mathcal{E} \mid \mathbf{y} = \mathbf{R}\mathbf{x} \text{ for all } \mathbf{x} \in [-1,1)^k\}.$$

2.2.1.4 Sublattices and cosets

Definition 2.2.1.6. *Let $\mathbf{V}_1$ and $\mathbf{V}_2$ be $k \times k$ integer matrices. $LAT(\mathbf{V}_2)$ is called a sublattice of $LAT(\mathbf{V}_1)$ if $LAT(\mathbf{V}_2) \subseteq LAT(\mathbf{V}_1)$, that is, every point of $LAT(\mathbf{V}_2)$ is also a point of $LAT(\mathbf{V}_1)$.*

Let $\mathbf{V}_1$ and $\mathbf{V}_2$ be integer matrices, where $LAT(\mathbf{V}_2) \subseteq LAT(\mathbf{V}_1)$. Then, for every $\mathbf{m} \in \mathcal{N}$, there exist $\mathbf{n} \in \mathcal{N}$ such that

$$\mathbf{V}_1\mathbf{n} = \mathbf{V}_2\mathbf{m}$$

or equivalently,

$$\mathbf{n} = \mathbf{V}_1^{-1}\mathbf{V}_2\mathbf{m}.$$

Since $\mathbf{n}$ and $\mathbf{m}$ are integer vectors and since $\mathbf{V}_1\mathbf{n} = \mathbf{V}_2\mathbf{m}$, then $\mathbf{V}_1^{-1}\mathbf{V}_2$ must be an integer-valued matrix. Let $\mathbf{L} = \mathbf{V}_1^{-1}\mathbf{V}_2$. Then, $\mathbf{V}_1\mathbf{L} = \mathbf{V}_2$. Since $\det \mathbf{V}_1\mathbf{L} = (\det \mathbf{V}_1)(\det \mathbf{L})$, then

$$\det \mathbf{V}_2 = (\det \mathbf{V}_1)(\det \mathbf{L})$$

so that,

$$J(\mathbf{V}_2) = J(\mathbf{V}_1)(\det \mathbf{L}).$$

Hence, $J(\mathbf{V}_2)$ is an integer multiple of $J(\mathbf{V}_1)$. For example, if $\mathbf{V}_2 = \begin{bmatrix} 4 & 0 \\ 0 & 4 \end{bmatrix}$ and $\mathbf{V}_1 = \begin{bmatrix} 2 & 0 \\ 0 & 2 \end{bmatrix}$, then $\mathbf{V}_2$ is a sublattice of $\mathbf{V}_1$. As such, $J(\mathbf{V}_2) = 16$ and $J(\mathbf{V}_1) = 4$, then $J(\mathbf{V}_2)$ is, as expected, an integer multiple of $J(\mathbf{V}_1)$, *i.e.* $J(\mathbf{V}_2) = 4J(\mathbf{V}_1)$. An important special case results when the lattice $LAT(\mathbf{V}_1)$ coincides with the fundamental lattice $\mathcal{N}$, *i.e.*, $LAT(\mathbf{V}_1) = LAT(\mathbf{I})$. In addition, $\rho = \frac{J(\mathbf{V}_2)}{J(\mathbf{V}_1)}$ represents the number of cells of $\text{FPD}(\mathbf{V}_1)$ which can fit into $\text{FPD}(\mathbf{V}_2)$. The lattice point in each one of these cells can be thought of as a shift vector **a**, which if added to each vector of LAT($\mathbf{V}_2$) will generate an equivalence class of points called a *coset*. The union of all cosets is LAT($\mathbf{V}_1$).

Thus, the concept of a coset will provide a natural way to partition LAT($\mathbf{V}_1$) into subsets, which is a necessary step for the generation of multidimensional polyphase components.

Definition 2.2.1.7. *Let* $\mathbf{V}_1$ *and* $\mathbf{V}_2$ *be integer-valued matrices such that*

$$LAT(\mathbf{V}_2) \subseteq LAT(\mathbf{V}_1).$$

Let $\mathbf{a} \in LAT(\mathbf{V}_1) \cap FPD(\mathbf{V}_2)$. *Define the coset*
$C(\mathbf{V}_1, \mathbf{V}_2, \mathbf{a})$ *to be*

$$C(\mathbf{V}_1, \mathbf{V}_2, \mathbf{a}) = LAT(\mathbf{V}_2) + \mathbf{a}.$$

If LAT($\mathbf{V}_1$) = $\mathcal{N}$, then by convention, $\mathbf{V}_1$ is not explicitly identified, that is, the coset is simply

$$C(\mathbf{V}_2, \mathbf{a}) = LAT(\mathbf{V}_2) + \mathbf{a}.$$

Given integer-valued matrices $\mathbf{V}_1$ and $\mathbf{V}_2$ such that LAT($\mathbf{V}_2$)$\subseteq$LAT($\mathbf{V}_1$), then denote the set of shift vectors by

$$\mathcal{N}(\mathbf{V}_1, \mathbf{V}_2) = \{\mathbf{a} \,|\, \mathbf{a} \in \mathrm{LAT}(\mathbf{V}_1) \cap \mathrm{FPD}(\mathbf{V}_2)\}.$$

Similar to the convention for cosets, the convention for shift vectors when LAT($\mathbf{V}_1$) = $\mathcal{N}$ is not to explicitly identify $\mathbf{V}_1$, that is, the set of shift vectors is simply

$$\mathcal{N}(\mathbf{V}_2) = \{\mathbf{a} \,|\, \mathbf{a} \in \mathcal{N} \cap \mathrm{FPD}(\mathbf{V}_2)\}.$$

Returning to the example above, if $\mathbf{V}_2 = \begin{bmatrix} 4 & 0 \\ 0 & 4 \end{bmatrix}$ and $\mathbf{V}_1 = \begin{bmatrix} 2 & 0 \\ 0 & 2 \end{bmatrix}$, then the cosets are uniquely defined by the following set of shift vectors

$$\mathcal{N}(\mathbf{V}_1, \mathbf{V}_2) = \left\{ \begin{bmatrix} 0 \\ 0 \end{bmatrix}, \begin{bmatrix} 2 \\ 0 \end{bmatrix}, \begin{bmatrix} 0 \\ 2 \end{bmatrix}, \begin{bmatrix} 2 \\ 2 \end{bmatrix} \right\}.$$

Let us briefly review some elementary operations that can be applied to integer matrices. These operations will be essential when we subsequently discuss the Smith form decomposition, a method for diagonalizing the sampling matrix.

2.2.1.5   Elementary operations

Elementary row (or column) operations on integer matrices are important because they permit the patterning of integer matrices into simpler forms, such as triangular and diagonal forms.

Definition 2.2.1.8. *Any elementary row operation on an integer-valued matrix* $\mathbf{P}$ *is defined to be any of the following:*

Type-1: *Interchange two rows.*

Type-2: *Multiply a row by a nonzero integer constant c.*

Type-3: *Add an integer multiple of a row to another row.*

These operations can be represented by premultiplying $\mathbf{P}$ with an appropriate square matrix called an elementary matrix. To illustrate these elementary operations, consider the following examples. (By convention, the rows and columns are numbered starting with zero rather than one.) The first example is a Type-1 elementary matrix that interchanges row 0 and row 3, which has the form

$$\begin{bmatrix} 0 & 0 & 0 & 1 \\ 0 & 1 & 0 & 0 \\ 0 & 0 & 1 & 0 \\ 1 & 0 & 0 & 0 \end{bmatrix}.$$

The second example is a Type-2 elementary matrix that multiplies elements in row 1 by $c \neq 0$, which has the form

$$\begin{bmatrix} 1 & 0 & 0 & 0 \\ 0 & c & 0 & 0 \\ 0 & 0 & 1 & 0 \\ 0 & 0 & 0 & 1 \end{bmatrix}.$$

The third example is a Type-3 elementary matrix that replaces row 3 with row 3 $+$ $(a * \text{row } 0)$, which has the form

$$\begin{bmatrix} 1 & 0 & 0 & 0 \\ 0 & 1 & 0 & 0 \\ 0 & 0 & 1 & 0 \\ a & 0 & 0 & 1 \end{bmatrix}.$$

All three types of elementary polynomial matrices are integer-valued unimodular matrices.

2.2.1.6   Smith form decomposition

The Smith Form Decomposition provides a method for diagonalizing the sampling matrix. When matrices are diagonal, most one-dimensional results can be extended automatically by performing operations in each dimension separately. However, in the nondiagonal case, these extensions are nontrivial and require complicated notations and matrix operations.

Theorem 2.2.1.5. *Every integer-valued matrix* **R** *can be expressed in its corresponding Smith form decomposition as*

$$\mathbf{R} = \mathbf{U}\mathbf{\Lambda}\mathbf{V}$$

where **U** *and* **V** *are integer-valued unimodular matrices and the Smith form* **Λ** *is given by*

$$\mathbf{\Lambda} = diag\,(\lambda_0, \ldots, \lambda_{r-1}, 0, 0, \ldots, 0)$$

where r is the rank of **R** *and* $\lambda_i |\ \lambda_{i+1}$, $i = 0, \ldots, r - 2$.

Proof: Assume the zeroth column of **R** contains a nonzero element, which may be brought to the $(0, 0)$ position by elementary operations. This element is the gcd of the zeroth column. If the new $(0, 0)$ element does not divide all the elements in the zeroth row, then it may be replaced by the gcd of the elements of the zeroth row (the effect will be that it will contain fewer prime factors than before). This process is repeated until an element in the $(0, 0)$ position is obtained which divides every element of the zeroth row and column. By elementary row and column operations, all the elements of the zeroth row and column, other than the $(0, 0)$ element, may be made zero. Denote the new submatrix formed by deleting the zeroth row and zeroth column by **C**.

Suppose that the submatrix **C** contains an element c_{ij} which is not divisible by c_{00}. Add column j to column 0. Column 0 then consists of the elements $c_{00}, c_{1,j}, \ldots, c_{n-1,j}$. Repeating the process, we replace c_{00} by a proper divisor of itself using elementary operations. Then, we must finally reach the stage where the element in the $(0, 0)$ position divides every element of the matrix, and all other elements of the zeroth row and column are zero.

The entire process is repeated with the submatrix obtained by deleting the zeroth row and column. Eventually a stage is reached when the matrix has the form

$$\begin{bmatrix} \mathbf{D} & \mathbf{0} \\ \mathbf{0} & \mathbf{E} \end{bmatrix}$$

where $\mathbf{D} = \text{diag}\,(\lambda_0, \ldots, \lambda_{r-1})$ and $\lambda_i | \lambda_{i+1}, i = 0, \ldots, r-2$. But $\mathbf{E}$ must be the zero matrix, since otherwise $\mathbf{R}$ would have rank larger than r.

■

Note that although the two unimodular matrices $\mathbf{U}$ and $\mathbf{V}$ are not unique, the diagonal matrix $\mathbf{\Lambda}$ is uniquely determined by $\mathbf{R}$.

Example 2.2.1.4. To illustrate the Smith form decomposition, consider the following example. Let

$$\mathbf{R} = \begin{bmatrix} 1 & 1 \\ 2 & -2 \end{bmatrix},$$

which corresponds to hexagonal sampling. If we divide the (1,0) element, 2, by the (0,0) element, 1, we obtain

$$2 = \underbrace{2}_{\text{quotient}}\,(1) + \underbrace{0}_{\text{remainder}}.$$

Therefore, if we apply a Type-3 row operation, which is defined by

$$\begin{bmatrix} 1 & 0 \\ -2 & 1 \end{bmatrix}$$

to $\mathbf{R}$, we will reduce the (1,0) element to zero. Therefore,

$$\begin{bmatrix} 1 & 0 \\ -2 & 1 \end{bmatrix} \begin{bmatrix} 1 & 1 \\ 2 & -2 \end{bmatrix} = \begin{bmatrix} 1 & 1 \\ 0 & -4 \end{bmatrix}.$$

Transform the (0,1) element to zero by a Type-3 column operation, which is defined by $\begin{bmatrix} 1 & -1 \\ 0 & 1 \end{bmatrix}$. Then, we obtain

$$\begin{bmatrix} 1 & 1 \\ 0 & -4 \end{bmatrix} \begin{bmatrix} 1 & -1 \\ 0 & 1 \end{bmatrix} = \begin{bmatrix} 1 & 0 \\ 0 & -4 \end{bmatrix}.$$

Finally, the (1,1) element is forced to be positive by a Type-2 row operation, which is defined by $\begin{bmatrix} 1 & 0 \\ 0 & -1 \end{bmatrix}$. Then, we obtain

$$\begin{bmatrix} 1 & 0 \\ 0 & -1 \end{bmatrix} \begin{bmatrix} 1 & 0 \\ 0 & -4 \end{bmatrix} = \begin{bmatrix} 1 & 0 \\ 0 & 4 \end{bmatrix}.$$

Thus,

$$\mathbf{\Lambda} = \begin{bmatrix} 1 & 0 \\ 0 & 4 \end{bmatrix}.$$

Let **E** be the product of elementary row operations, *i.e.*

$$\mathbf{E} = \begin{bmatrix} 1 & 0 \\ 0 & -1 \end{bmatrix} \begin{bmatrix} 1 & 0 \\ -2 & 1 \end{bmatrix} = \begin{bmatrix} 1 & 0 \\ 2 & -1 \end{bmatrix}.$$

Let **F** be the product of elementary column operations, *i.e.*

$$\mathbf{F} = \begin{bmatrix} 1 & -1 \\ 0 & 1 \end{bmatrix},$$

since only one elementary column operation was performed. Therefore,

$$\mathbf{ERF} = \mathbf{\Lambda}.$$

Then, the Smith form decomposition is given by

$$\mathbf{R} = \mathbf{U\Lambda V}$$

where,

$$\mathbf{U} = \mathbf{E}^{-1} = \begin{bmatrix} 1 & 0 \\ 2 & -1 \end{bmatrix},$$

$$\mathbf{\Lambda} = \begin{bmatrix} 1 & 0 \\ 0 & 4 \end{bmatrix},$$

and

$$\mathbf{V} = \mathbf{F}^{-1} = \begin{bmatrix} 1 & 1 \\ 0 & 1 \end{bmatrix}.$$

Theorem 2.2.1.6. *Let* **U** *and* **V** *be unimodular matrices and let* **Λ** *be a diagonal matrix. If the Smith form decomposition of sampling matrix* **R** *is given by* **R** = **UΛV**, *then*

$$FPD(\mathbf{R}) = \mathbf{U} \ FPD(\mathbf{\Lambda}).$$

Proof: Since LAT(**ΛV**) = LAT(**Λ**) by Theorem 2.2.1.2, then FPD(**ΛV**) = FPD(**Λ**). Therefore, FPD(**UΛV**) = FPD(**UΛ**). Since FPD(**UΛ**) = **U** FPD(**Λ**) by Theorem 2.2.1.4, then FPD(**R**) = **U** FPD(**Λ**). ∎

2.2.2 Multidimensional sampled signals

Unfortunately, some important sampling structures can not be represented as a lattice. For example, consider an important sampling structure for High Definition Television(HDTV) called *line quincunx*, where two samples are placed one vertically above the other in place of every sample in the sampling grid. But, line quincunx can be represented as the union of two shifted lattices using the multidimensional z-transform. We will first define some underlying vector mathematics and, then we will present the multidimensional z-transform and the multidimensional discrete Fourier transform. Then we will present two multidimensional signal representations — the modulation representation and the polyphase representation. The theory of multirate and wavelet signal processing is considerably simplified by the use of these representations.

2.2.2.1   Vector mathematics

In order to generalize the definitions that we have grown accustomed to seeing in one-dimension, we will provide the definition of a vector raised to a vector power and, subsequently, the definition of a vector raised to a matrix power.

Definition 2.2.2.1. *Given complex-valued vector* $\mathbf{r} = \begin{bmatrix} r_0, & \ldots, & r_{N-1} \end{bmatrix}^T$ *and integer-valued vector* $\mathbf{s} = \begin{bmatrix} s_0, & \ldots, & s_{N-1} \end{bmatrix}^T$. *Then, the vector* $\mathbf{r}$ *raised to the vector* $\mathbf{s}$ *power is a scalar and it is defined to be*

$$\mathbf{r}^\mathbf{s} = r_0^{s_0} \; r_1^{s_1} \; \ldots \; r_{N-1}^{s_{N-1}},$$

or equivalently,

$$\mathbf{r}^\mathbf{s} = \prod_{m=0}^{N-1} r_m^{s_m}.$$

Then, building on this definition, we will define a vector raised to a matrix power.

Definition 2.2.2.2. *Given a complex-valued vector* $\mathbf{r} = \begin{bmatrix} r_0, & \ldots, & r_{N-1} \end{bmatrix}^T$ *and an integer-valued matrix* $\mathbf{L} = \begin{bmatrix} L_0, & \ldots, & L_{N-1} \end{bmatrix}$, *where* L_i *is the* i*th column of* $\mathbf{L}$. *Then, the vector* $\mathbf{r}$ *raised to the matrix* $\mathbf{L}$ *power is a row vector and it is defined to be*

$$\mathbf{r}^\mathbf{L} = \begin{bmatrix} \mathbf{r}^{L_0}, & \mathbf{r}^{L_1}, & \ldots, & \mathbf{r}^{L_{N-1}} \end{bmatrix}.$$

2.2.2.2   Multidimensional z-transform

Definition 2.2.2.3. *The* k*-dimensional* z*-transform of* $x(n_0, \ldots, n_{k-1})$ *is defined by*

$$X(z_0, \ldots, z_{k-1}) = \sum_{n_0} \cdots \sum_{n_{k-1}} x(n_0, \ldots, n_{k-1}) \, z_0^{-n_0} \ldots z_{k-1}^{-n_{k-1}},$$

or equivalently,

$$X(\mathbf{z}) = \sum_{\mathbf{n} \in \mathcal{N}} x(\mathbf{n}) \mathbf{z}^{-\mathbf{n}}$$

where, $\mathbf{z} = [z_0, ..., z_{k-1}]^T$ *is a complex-valued vector,* $\mathbf{n} = [n_0, ..., n_{k-1}]^T$ *is an integer-valued vector, and*

$$\mathbf{z}^{-\mathbf{n}} = \prod_{m=0}^{k-1} z_m^{-n_m}.$$

Let $\mathbf{L}$ *be an integer-valued nonsingular matrix, then by Definition 2.2.2.2,* $\mathbf{z}^{\mathbf{L}}$ *is given by*

$$\mathbf{z}^{\mathbf{L}} = \left[\mathbf{z}^{L_0}, \dots, \mathbf{z}^{L_{k-1}} \right]$$

where, L_i *is the* i*th column of* $\mathbf{L}$*, that is,*

$$\mathbf{L} = [L_0, \dots, L_{k-1}].$$

Theorem 2.2.2.1. *Let* $\mathbf{L}$ *be an integer-valued matrix where* $L_i, i = 0, \dots, k-1$*, are the columns of* $\mathbf{L}$*. Then,*

$$\left(\mathbf{z}^{\mathbf{L}} \right)^{\mathbf{n}} = \mathbf{z}^{\mathbf{L}\mathbf{n}}.$$

Proof: Using the Definition 2.2.2.2, we can write $\left(\mathbf{z}^{\mathbf{L}} \right)^{\mathbf{n}}$ as

$$\left(\mathbf{z}^{\mathbf{L}} \right)^{\mathbf{n}} = \left[\mathbf{z}^{L_0}, \dots, \mathbf{z}^{L_{k-1}} \right]^{\mathbf{n}}$$

where, L_i is the ith column of $\mathbf{L}$. Substituting the definition of a vector raised to a vector power yields

$$\left(\mathbf{z}^{\mathbf{L}} \right)^{\mathbf{n}} = \left[\prod_{m=0}^{k-1} z_m^{L_{m,0}}, \dots, \prod_{m=0}^{k-1} z_m^{L_{m,k-1}} \right]^{\mathbf{n}},$$

where $L_{m,i}$ is the mth component of L_i. By definition, a vector raised to a vector power gives a scalar. Hence,

$$\left(\mathbf{z}^{\mathbf{L}} \right)^{\mathbf{n}} = \prod_{p=0}^{k-1} \left(\prod_{m=0}^{k-1} z_m^{(L_{m,p})(n_p)} \right),$$

or equivalently,

$$\left(\mathbf{z}^{\mathbf{L}} \right)^{\mathbf{n}} = \prod_{m=0}^{k-1} \left(\prod_{p=0}^{k-1} z_m^{(L_{m,p})(n_p)} \right).$$

Since the product of terms with the same base equals the base to the sum of the exponents, the last equation becomes

$$\left(\mathbf{z}^{\mathbf{L}}\right)^{\mathbf{n}} = \prod_{m=0}^{k-1} z_m^{\sum_{p=0}^{k-1}(L_{m,p})(n_p)},$$

or equivalently,

$$\left(\mathbf{z}^{\mathbf{L}}\right)^{\mathbf{n}} = \mathbf{z}^{\mathbf{Ln}}.$$

∎

If the z-transform converges for all z_m of the form $z_m = \exp(j\omega_m)$, $m = 0, \ldots, k-1$, then the z-transform can be represented as the sum of harmonically related sinusoids, i.e.

$$X(\underline{\omega}) = \sum_{\mathbf{n} \in \mathcal{N}} x(\mathbf{n}) \exp(-j\underline{\omega}^T \mathbf{n}),$$

which is the multidimensional generalization of the discrete-time Fourier transform. In order to quickly distinguish vectors in Fourier space, the vector Fourier variable $\underline{\omega}$ is denoted by an underlined omega rather than a bold omega.

Theorem 2.2.2.2. *Let* **L** *be an integer-valued matrix. Then,*

$$\exp\left(j\underline{\omega}\right)^{\mathbf{L}} = \exp\left(j\mathbf{L}^T\underline{\omega}\right).$$

Proof: Using the Definition 2.2.2.2, we find that

$$\exp(j\underline{\omega})^{\mathbf{L}} = \left[\exp(j\underline{\omega}^T L_0), \ldots, \exp(j\underline{\omega}^T L_{k-1})\right]$$

where, L_i is the ith column of **L**. Since the exponents of each term are simply inner products between $\underline{\omega}$ and a column of **L**, their order can be interchanged, that is,

$$\exp(j\underline{\omega})^{\mathbf{L}} = \left[\exp(jL_0^T\underline{\omega}), \ldots, \exp(jL_{k-1}^T\underline{\omega})\right],$$

or equivalently,

$$\exp(j\underline{\omega})^{\mathbf{L}} = \exp(j\mathbf{L}^T\underline{\omega}).$$

∎

2.2.2.3 Multidimensional discrete Fourier transform

The multidimensional discrete Fourier transform is an exact Fourier representation for periodically sampled arrays. Therefore, it takes the form of a periodically sampled Fourier transform. As in the one-dimensional case, the multidimensional discrete Fourier transform can be interpreted as a Fourier series representation for one period of a periodic sequence.

In this formulation, we will have to address two types of periodicities – one due to the sampling lattice and one due to the signal (that is, defined on lattice points) to be Fourier transformed. Let **V** denote the sampling matrix, *i.e.*, hexagonal, quincunx, rectangular, etc. Let **N** denote the periodicity matrix, which characterizes the periodicity of the lattice points on which the signal to be Fourier transformed is defined. Assume LAT(**N**) is a sublattice of LAT(**V**). Then, we define equivalence classes between periodic replicas of the data by

$$[\mathbf{n}] = \{\ \mathbf{m} \in \text{LAT}(\mathbf{V}) \mid \mathbf{n} - \mathbf{m} \in \text{LAT}(\mathbf{N})\ \}.$$

Therefore, if parallelograms are drawn between the elements of LAT(**N**), then any two vectors that occupy the same relative position are in the same equivalence class.

Many properties of the periodicity matrix, **N**, follow by analogy from the corresponding facts for sampling matrices. For example, the density of the periodicity matrix is uniquely defined by |det **N** |, denoted J(**N**); but for a given periodic sequence the periodicity matrix **N** is not unique, since it can be multiplied by any unimodular matrix and still describe the same periodic signal. In addition, the columns of **N** indicate the vectors along which it is periodically replicated.

Definition 2.2.2.4. *A multidimensional sequence* $x(\mathbf{n})$ *is periodic with period* **N**, *that is, for all* $\mathbf{n}, \mathbf{r} \in \mathcal{N}$, $x(\mathbf{n}) = x(\mathbf{n} + \mathbf{N}\mathbf{r})$.

Let $\mathcal{I}_\mathbf{N}$ represent one period of $x(\mathbf{n})$. Then,

$$X(\underline{\omega}) = \sum_{\mathbf{n} \in \mathcal{I}_\mathbf{N}} x(\mathbf{n}) \exp[-j\underline{\omega}^T \mathbf{V}\mathbf{n}]$$

where **V** defines the underlying sampling lattice. Moreover, since $x(\mathbf{n})$ is periodic with period **N**, $X(\underline{\omega})$ can also be written as

$$X(\underline{\omega}) = \sum_{\mathbf{n} \in \mathcal{I}_\mathbf{N}} x(\mathbf{n}) \exp[-j\underline{\omega}^T \mathbf{V}(\mathbf{n} + \mathbf{N}\mathbf{r})],$$

or simply

$$X(\underline{\omega}) = \sum_{\mathbf{n} \in \mathcal{I}_\mathbf{N}} x(\mathbf{n}) \exp[-j\underline{\omega}^T \mathbf{V}\mathbf{n}] \exp[-j\underline{\omega}^T \mathbf{V}\mathbf{N}\mathbf{r}].$$

But,

$$X(\underline{\omega}) = \sum_{\mathbf{n}_i \in \mathcal{I}_\mathbf{N}} x(\mathbf{n}) \exp[-j\underline{\omega}^T \mathbf{V}\mathbf{n}]\ .$$

Therefore,

$$\exp[-j\underline{\omega}^T \mathbf{V} \mathbf{N} \mathbf{r}] = 1,$$

which is equivalent to the condition,

$$\underline{\omega}^T \mathbf{V} \mathbf{N} = 2\pi \mathbf{m}^T,$$

where $\mathbf{m}$ is a vector of integers. Upon further examination of $\underline{\omega}^T$, we observe that

$$\underline{\omega}^T \;=\; 2\pi \mathbf{m}^T (\mathbf{V} \mathbf{N})^{-1},$$

or equivalently,

$$\underline{\omega}^T \;=\; \left(2\pi (\mathbf{V} \mathbf{N})^{-T} \mathbf{m}\right)^T.$$

Therefore,

$$\underline{\omega} = 2\pi (\mathbf{V} \mathbf{N})^{-T} \mathbf{m}.$$

The matrix $2\pi (\mathbf{V} \mathbf{N})^{-T}$ serves as a Fourier domain sampling matrix. Substituting this equation into the equation for $X(\underline{\omega})$ yields

$$X(\mathbf{m}) = \sum_{\mathbf{n} \in \mathcal{I}_{\mathbf{N}}} x(\mathbf{n}) \exp[-j 2\pi \mathbf{m}^T \mathbf{N}^{-1} \mathbf{V}^{-1} \mathbf{V} \mathbf{n}],$$

or equivalently,

$$X(\mathbf{m}) = \sum_{\mathbf{n} \in \mathcal{I}_{\mathbf{N}}} x(\mathbf{n}) \exp[-j \mathbf{m}^T (2\pi \mathbf{N}^{-1}) \mathbf{n}].$$

Let us further examine the inner product which occurs in the argument of the exponential.

$$\begin{aligned}
\mathbf{m}^T (2\pi \mathbf{N})^{-1} \mathbf{n} &= \left(2\pi \mathbf{N}^{-1} \mathbf{n}\right)^T \mathbf{m} \\
&= \mathbf{n}^T (2\pi \mathbf{N}^{-T}) \mathbf{m}.
\end{aligned}$$

Therefore,

$$X(\mathbf{m}) = \sum_{\mathbf{n} \in \mathcal{I}_{\mathbf{N}}} x(\mathbf{n}) \exp[-j \mathbf{n}^T (2\pi \mathbf{N}^{-T}) \mathbf{m}].$$

Suppose the multidimensional sequence $X(\mathbf{m})$ is periodic with period $\mathbf{P}$, that is, $X(\mathbf{m}) = X(\mathbf{m} + \mathbf{P} \mathbf{q})$ for $\mathbf{m}, \mathbf{q} \in \mathcal{N}$. Also, let $\mathcal{I}_{\mathbf{P}}$ represent one period of $X(\mathbf{m})$. Then, by analogy with the one-dimensional discrete Fourier transform, assume that $x(\mathbf{n})$ has the following form for some constant α,

$$x(\mathbf{n}) = \frac{1}{\alpha} \sum_{\mathbf{m} \in \mathcal{I}_{\mathbf{P}}} X(\mathbf{m}) \exp[j \mathbf{n}^T (2\pi \mathbf{N}^{-T}) \mathbf{m}].$$

Invoking the periodicity of $X(\mathbf{m})$, that is, $X(\mathbf{m}) = X(\mathbf{m} + \mathbf{Pq})$, will cause $x(\mathbf{n})$ to become

$$x(\mathbf{n}) = \frac{1}{\alpha} \sum_{\mathbf{m} \in \mathcal{I}_\mathbf{P}} X(\mathbf{m}) \exp[j\mathbf{n}^T (2\pi\mathbf{N}^{-T})(\mathbf{m} + \mathbf{Pq})],$$

or equivalently,

$$x(\mathbf{n}) = \frac{1}{\alpha} \sum_{\mathbf{m} \in \mathcal{I}_\mathbf{P}} X(\mathbf{m}) \exp[j\mathbf{n}^T (2\pi\mathbf{N}^{-T})\mathbf{m}] \exp[j\mathbf{n}^T (2\pi\mathbf{N}^{-T})\mathbf{Pq}].$$

But,

$$x(\mathbf{n}) = \frac{1}{\alpha} \sum_{\mathbf{m} \in \mathcal{I}_\mathbf{P}} X(\mathbf{m}) \exp[j\mathbf{n}^T (2\pi\mathbf{N}^{-T})\mathbf{m}].$$

Therefore,

$$\exp[j\mathbf{n}^T (2\pi\mathbf{N}^{-T})\mathbf{Pq}] = 1 \text{ for all } \mathbf{q} \in \mathcal{N}.$$

Since $\mathbf{n}$ and $\mathbf{q}$ are integer-valued vectors, then

$$\mathbf{N}^{-T}\mathbf{P} = \mathbf{I},$$

or equivalently,

$$\mathbf{P} = \mathbf{N}^T.$$

Therefore, $X(\mathbf{m})$ is periodic with period $\mathbf{N}^T$, that is, $X(\mathbf{m}) = X(\mathbf{m} + \mathbf{N}^T\mathbf{q})$. Hence,

$$x(\mathbf{n}) = \frac{1}{\alpha} \sum_{\mathbf{m} \in \mathcal{I}_{\mathbf{N}^T}} X(\mathbf{m}) \exp[j\mathbf{n}^T (2\pi\mathbf{N}^{-T})\mathbf{m}].$$

Now let us determine the constant α by substituting the equation for $x(\mathbf{n})$ into the equation for $X(\mathbf{m})$. Hence,

$$X(\mathbf{m}) = \frac{1}{\alpha} \sum_{\mathbf{s} \in \mathcal{I}_{\mathbf{N}^T}} X(\mathbf{s}) \sum_{\mathbf{n} \in \mathcal{I}_\mathbf{N}} \exp[j\mathbf{n}^T (2\pi\mathbf{N}^{-T})\mathbf{m}] \exp[-j\mathbf{n}^T (2\pi\mathbf{N}^{-T})\mathbf{s}],$$

or equivalently,

$$X(\mathbf{m}) = \frac{1}{\alpha} \sum_{\mathbf{s} \in \mathcal{I}_{\mathbf{N}^T}} X(\mathbf{s}) \sum_{\mathbf{n} \in \mathcal{I}_\mathbf{N}} \exp[j\mathbf{n}^T (2\pi\mathbf{N}^{-T})(\mathbf{m} - \mathbf{s})].$$

However,

$$\sum_{\mathbf{n} \in \mathcal{I}_\mathbf{N}} \exp[j\mathbf{n}^T (2\pi\mathbf{N}^{-T})(\mathbf{m} - \mathbf{s})] = \mathbf{J}(\mathbf{N})\delta_{\mathbf{m},\mathbf{s}}.$$

Hence,

$$\alpha = \mathbf{J}(\mathbf{N})$$

which is as expected, since $J(N) = |\det N|$ is the number of samples in one period for $LAT(N)$. Therefore, the multidimensional discrete Fourier transform pair are given by:

$$X(\mathbf{m}) = \sum_{\mathbf{n} \in \mathcal{I}_N} x(\mathbf{n}) \exp[-j\mathbf{n}^T (2\pi N^{-T})\mathbf{m}],$$

and

$$x(\mathbf{n}) = \frac{1}{J(N)} \sum_{\mathbf{m} \in \mathcal{I}_{N^T}} X(\mathbf{m}) \exp[j\mathbf{n}^T (2\pi N^{-T})\mathbf{m}].$$

It should be noted that these equations reduce to the usual discrete Fourier transform pair in the one-dimensional case and to the familiar rectangular multidimensional discrete Fourier transform when N is a diagonal matrix.

As an illustration of this theoretical development, sometimes it is of interest to input data from an arbitrary lattice and output it on a rectangular lattice, so that it could be conveniently displayed on a computer display. Assume that V is defined by

$$V = \begin{bmatrix} a & b \\ 0 & c \end{bmatrix}.$$

For hexagonal input: $a = 2$, $b = 1$, $c = 2$. Moreover, for quincunx input: $a = 2$, $b = 1$, $c = 1$. In addition, for rectangular input: $b = 0$. Select a periodicity matrix so that VN is a diagonal matrix so that the resulting Fourier analysis will be on a rectangular grid. Now, let us pick N to be

$$N = \begin{bmatrix} N_1 & -\frac{b}{a}N_2 \\ 0 & N_2 \end{bmatrix}.$$

Then,

$$VN = \begin{bmatrix} a & b \\ 0 & c \end{bmatrix} \begin{bmatrix} N_1 & -\frac{b}{a}N_2 \\ 0 & N_2 \end{bmatrix} = \begin{bmatrix} aN_1 & 0 \\ 0 & cN_2 \end{bmatrix},$$

then this N matrix is a good choice for a periodicity matrix. Therefore, the DFT becomes

$$X(\mathbf{m}) = \sum_{\mathbf{n} \in \mathcal{I}_N} x(\mathbf{n}) \exp[-j\mathbf{n}^T (2\pi N^{-T})\mathbf{m}]$$

where,

$$N^{-T} = \begin{bmatrix} \frac{1}{N_1} & 0 \\ \frac{b}{a\,N_1} & \frac{1}{N_2} \end{bmatrix}.$$

This suggests the following algorithm:

(1) Compute $N_2 N_1$-point FFTs, one for each row in the n_1 direction.

$$X_1(m_1, n_2) = \sum_{n_1=0}^{N_1-1} x(n_1, n_2) \exp\left(-j2\pi\frac{n_1 m_1}{N_1}\right).$$

(2) Apply a phase shift to each point of the resulting data.

$$X_2(m_1, n_2) = X_1(m_1, n_2) \exp\left(-j\frac{2\pi}{N_1} \, \text{round}\!\left(\frac{b}{a} n_2 m_1\right)\right).$$

Since we are working with a sampling grid with samples at integer-valued locations, it is important that we perform the phase shift for integer multiples of $\frac{2\pi}{N_1}$. But $\frac{b}{a} n_2 m_1$ is real-valued. Therefore, we will need to quantize $\frac{b}{a} n_2 m_1$ to integer values through the use of the round function.

(3) Compute $N_1 N_2$-point FFTs, one for each column in the n_2 direction.

$$X_3(m_1, m_2) = \sum_{n_2=0}^{N_2-1} X_2(m_1, n_2) \exp\left(-j2\pi\frac{n_2 m_2}{N_2}\right).$$

2.2.2.4 The Smith form and the DFT

First, let us begin with the multidimensional discrete Fourier transform, that is,

$$F(\mathbf{k}) = \sum_{\mathbf{n} \in \mathcal{I}_\mathbf{M}} f(\mathbf{n}) \exp\left[-j2\pi\mathbf{n}^T\left(\mathbf{M}^{-T}\mathbf{k}\right)\right].$$

Replace $\mathbf{M}$ with its Smith form decomposition, that is, $\mathbf{M} = \mathbf{U}\mathbf{\Lambda}\mathbf{V}$. Then, using Theorem 2.2.1.2, LAT($\mathbf{U}\mathbf{\Lambda}\mathbf{V}$) = LAT($\mathbf{U}\mathbf{\Lambda}$). Now, using elementary linear algebraic operations, let us simplify the exponent of the exponential

$$
\begin{aligned}
\exp\left[-j2\pi\mathbf{n}^T(\mathbf{M}^{-T}\mathbf{k})\right] &= \exp\left[-j2\pi\mathbf{n}^T(\mathbf{U}\mathbf{\Lambda}\mathbf{V})^{-T}\mathbf{k}\right] \\
&= \exp\left[-j2\pi((\mathbf{U}\mathbf{\Lambda}\mathbf{V})^{-1}\mathbf{n})^T\mathbf{k}\right] \\
&= \exp\left[-j2\pi\mathbf{k}^T(\mathbf{U}\mathbf{\Lambda}\mathbf{V})^{-1}\mathbf{n}\right] \\
&= \exp\left[-j2\pi\left(\mathbf{k}^T\mathbf{V}^{-1}\right)\mathbf{\Lambda}^{-1}\left(\mathbf{U}^{-1}\mathbf{n}\right)\right].
\end{aligned}
$$

Then, the multidimensional DFT becomes

$$F(\mathbf{V}^T\mathbf{p}) = \sum_{\mathbf{m} \in \mathcal{I}_\mathbf{\Lambda}} f(\mathbf{U}\mathbf{m}) \exp\left[-j2\pi\mathbf{p}^T\mathbf{\Lambda}^{-1}\mathbf{m}\right]$$

where

$$\mathbf{m} = \mathbf{U}^{-1}\mathbf{n} \text{ and } \mathbf{p}^T = \mathbf{k}^T\mathbf{V}^{-1}.$$

Therefore, the Smith form permits the use of a rectangular DFT, when the initial data lies on a nonrectangular sampling grid. The initial data must have parallelepiped spatial support which becomes rectangular after being mapped by U^{-1}. Moreover, the larger the values of the elements of U^{-1}, the more the spatial support will be skewed.

Thus, the algorithm for the Smith form version of the multidimensional DFT is given by the following:

(1) Shuffle the input data samples by U^{-1}.

(2) Perform a separable multidimensional DFT with lengths equal to the diagonal elements of Λ.

(3) Shuffle the output data samples by V^T.

Let us examine more carefully the mapping between the space of the initial data samples and the space defined by Λ. Assume that the data is defined on a quincunx grid with the following Smith form decomposition.

$$\underbrace{\begin{bmatrix} 1 & 1 \\ -1 & 1 \end{bmatrix}}_{M} = \underbrace{\begin{bmatrix} 1 & 0 \\ -1 & 1 \end{bmatrix}}_{U} \underbrace{\begin{bmatrix} 1 & 0 \\ 0 & 2 \end{bmatrix}}_{\Lambda} \underbrace{\begin{bmatrix} 1 & 1 \\ 0 & 1 \end{bmatrix}}_{V}.$$

Then, Λ characterizes the intermediate space for the Smith Form decomposition and the mapping from M to Λ can be visualized by Figures 2.6 and 2.7.

Similarly, the inverse DFT is given by

$$f(Um) = \frac{1}{J(\Lambda)} \sum_{p \in \mathcal{I}_\Lambda} F(V^T p) \exp \left[j 2\pi p^T \Lambda^{-1} m \right]$$

where,

$$J(M) = | \det M | = | \det U \| \det \Lambda \| \det V | = | \det \Lambda | = J(\Lambda).$$

2.2.2.5   Modulation representation

Definition 2.2.2.5. *Let M be the sampling matrix. Then, given a multidimensional sequence $x(n)$, the components of the multidimensional modulation representation of the multidimensional z-transform of $x(n)$ are defined by*

$$X_h^{(M)}(z) = X \left[z \, \exp[-j(2\pi M^{-T})h] \right] ; \quad h \in \mathcal{I}_{M^T}.$$

To more easily interpret this equation, assume z takes on the value of $\exp(j\underline{\omega})$. Then,

$$X_h^{(M)}(\exp(j\underline{\omega})) = X \left(\exp[j\underline{\omega} - j(2\pi M^{-T})h] \right).$$

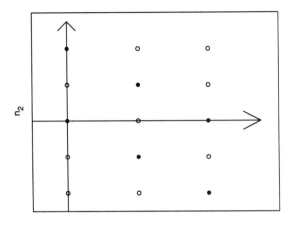

Figure 2.6. Input data.

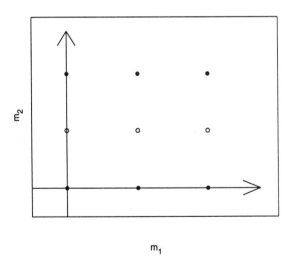

Figure 2.7. After shuffling input data.

But $z_m = \exp(j\omega_m)$ for $m = 0, \ldots, k-1$. Therefore,

$$X_\mathbf{h}^{(\mathbf{M})}(\mathbf{z}) = X\left[\exp[-j(2\pi\mathbf{M}^{-T})\mathbf{h}_0]\, z_0, \ldots, \exp[-j(2\pi\mathbf{M}^{-T})\mathbf{h}_{k-1}]\, z_{k-1}\right].$$

Example 2.2.2.1. This example illustrates the idea of the multidimensional modulation representation of $X(\mathbf{z})$ with respect to a rectangular sampling matrix $\mathbf{M} = \begin{bmatrix} 2 & 0 \\ 0 & 2 \end{bmatrix}$. In this case, $\mathbf{I}_{\mathbf{M}^T}$ will contain four values, that is, $\mathbf{h}_0 = \begin{bmatrix} 0 & 0 \end{bmatrix}^T$, $\mathbf{h}_1 = \begin{bmatrix} 0 & 1 \end{bmatrix}^T$, $\mathbf{h}_2 = \begin{bmatrix} 1 & 0 \end{bmatrix}^T$, and $\mathbf{h}_3 = \begin{bmatrix} 1 & 1 \end{bmatrix}^T$. Then,

$$X_{\mathbf{h}_l}^{(\mathbf{M})}(\mathbf{z}) = \begin{cases} X(z_0, z_1) & \text{for} \quad \mathbf{h}_0 = \begin{bmatrix} 0 & 0 \end{bmatrix}^T \\ X(z_0, -z_1) & \text{for} \quad \mathbf{h}_1 = \begin{bmatrix} 0 & 1 \end{bmatrix}^T \\ X(-z_0, z_1) & \text{for} \quad \mathbf{h}_2 = \begin{bmatrix} 1 & 0 \end{bmatrix}^T \\ X(-z_0, -z_1) & \text{for} \quad \mathbf{h}_3 = \begin{bmatrix} 1 & 1 \end{bmatrix}^T. \end{cases}$$

2.2.2.6   Polyphase representation

Definition 2.2.2.6. *Given a multidimensional sequence $x(\mathbf{n})$ and a nonsingular matrix $\mathbf{R}$, then its Type-I multidimensional polyphase components are defined as $x_\mathbf{a}(\mathbf{n}) = x(\mathbf{Rn} + \mathbf{a})$, where $\mathbf{n} \in \mathcal{N}$ and $\mathbf{a} \in \mathcal{N}(\mathbf{R})$.*

Now, let us investigate the multidimensional z-transform of the Type-I multidimensional polyphase components, that is,

$$X(\mathbf{z}) = \sum_{\mathbf{a} \in \mathcal{N}(\mathbf{R})} \sum_{\mathbf{n} \in \mathcal{N}} x(\mathbf{Rn} + \mathbf{a})\mathbf{z}^{-(\mathbf{Rn}+\mathbf{a})},$$

or equivalently,

$$X(\mathbf{z}) = \sum_{\mathbf{a} \in \mathcal{N}(\mathbf{R})} \mathbf{z}^{-\mathbf{a}} \sum_{\mathbf{n} \in \mathcal{N}} x(\mathbf{Rn} + \mathbf{a})(\mathbf{z}^\mathbf{R})^{-\mathbf{n}}.$$

For notational purposes, let

$$X_\mathbf{a}^{(\mathbf{R})}(\mathbf{z}) = \sum_{\mathbf{n} \in \mathcal{N}} x(\mathbf{Rn} + \mathbf{a})\mathbf{z}^{-\mathbf{n}} = \sum_{\mathbf{n} \in \mathcal{N}} x_\mathbf{a}(\mathbf{n})\mathbf{z}^{-\mathbf{n}},$$

then, $X(\mathbf{z})$ becomes

$$X(\mathbf{z}) = \sum_{\mathbf{a} \in \mathcal{N}(\mathbf{R})} \mathbf{z}^{-\mathbf{a}} X_\mathbf{a}^{(\mathbf{R})}(\mathbf{z}^\mathbf{R}),$$

where $\mathbf{z}^{-\mathbf{a}}$ corresponds to shifts of the multidimensional signal by a vector amount $\mathbf{a}$. Now consider the following example, which illustrates the idea of a Type-I multidimensional polyphase decomposition of the multidimensional filter $X(\mathbf{z})$.

Example 2.2.2.2. What is the Type-I multidimensional polyphase decomposition of the multidimensional filter $X(\mathbf{z})$ with respect to a rectangular sampling matrix $\mathbf{R} = \begin{bmatrix} 2 & 0 \\ 0 & 2 \end{bmatrix}$? In this case, there are $\mathrm{J}(\mathbf{R}) = |\det \mathbf{R}| = 4$ values of the shift vectors are given by $[0,0]^T$, $[1,0]^T$, $[0,1]^T$, and $[1,1]^T$. Let $\mathbf{z} = [z_0, z_1]^T$ and $\mathbf{a} = [(\mathbf{a})_0, (\mathbf{a})_1]^T$ then, $\mathbf{z}^{-\mathbf{a}} = z_0^{-(\mathbf{a})_0} z_1^{-(\mathbf{a})_1}$. Thus, the quantity $\mathbf{z}^{-\mathbf{a}}$ is given by

$$\mathbf{z}^{-\mathbf{a}} = \begin{cases} z_0^{-0} z_1^{-0} & \text{for} \quad \mathbf{a} = [0,0]^T \\ z_0^{-1} z_1^{-0} & \text{for} \quad \mathbf{a} = [1,0]^T \\ z_0^{-0} z_1^{-1} & \text{for} \quad \mathbf{a} = [0,1]^T \\ z_0^{-1} z_1^{-1} & \text{for} \quad \mathbf{a} = [1,1]^T, \end{cases}$$

or simply,

$$\mathbf{z}^{-\mathbf{a}} = \begin{cases} 1 & \text{for} \quad \mathbf{a} = [0,0]^T \\ z_0^{-1} & \text{for} \quad \mathbf{a} = [1,0]^T \\ z_1^{-1} & \text{for} \quad \mathbf{a} = [0,1]^T \\ z_0^{-1} z_1^{-1} & \text{for} \quad \mathbf{a} = [1,1]^T. \end{cases}$$

Since for arbitrary $\mathbf{R} = \begin{bmatrix} x_0 & x_1 \\ y_0 & y_1 \end{bmatrix}$, $\mathbf{z}^{\mathbf{R}}$ is given by $\mathbf{z}^{\mathbf{R}} = \left(z_0^{x_0} z_1^{y_0}, z_0^{x_1} z_1^{y_1} \right)$.
Hence, the quantity $X_{\mathbf{a}}^{(\mathbf{R})}(\mathbf{z}^{\mathbf{R}})$ is defined by

$$X_{\mathbf{a}}^{(\mathbf{R})}(\mathbf{z}^{\mathbf{R}}) = X_{\mathbf{a}}^{(\mathbf{R})}(z_0^2 z_1^0, z_0^0 z_1^2) = X_{\mathbf{a}}^{(\mathbf{R})}(z_0^2, z_1^2).$$

So, the polyphase decomposition of $X(\mathbf{z})$ is given by

$$\begin{aligned} X(\mathbf{z}) = & \; X_{\mathbf{a}}^{(\mathbf{R})}(z_0^2, z_1^2) + z_0^{-1} X_{\mathbf{a}}^{(\mathbf{R})}(z_0^2, z_1^2) \\ & + z_1^{-1} X_{\mathbf{a}}^{(\mathbf{R})}(z_0^2, z_1^2) + z_0^{-1} z_1^{-1} X_{\mathbf{a}}^{(\mathbf{R})}(z_0^2, z_1^2). \end{aligned}$$

Definition 2.2.2.7. *Given a multidimensional vector $x(\mathbf{n})$ and a nonsingular matrix $\mathbf{R}$, its Type-II multidimensional polyphase components are given by $x_{\mathbf{a}}(\mathbf{n}) = x(\mathbf{R}\mathbf{n} - \mathbf{a})$, where $\mathbf{a} \in \mathcal{N}(\mathbf{R})$ and $\mathbf{n} \in \mathcal{N}$.*

Now, let us investigate the multidimensional z-transform of the Type-II multidimensional polyphase components, that is,

$$X(z) = \sum_{\mathbf{a} \in \mathcal{N}(\mathbf{R})} \sum_{\mathbf{n} \in \mathcal{N}} x(\mathbf{Rn} - \mathbf{a})\mathbf{z}^{-(\mathbf{Rn}-\mathbf{a})},$$

or equivalently,

$$X(z) = \sum_{\mathbf{a} \in \mathcal{N}(\mathbf{R})} \mathbf{z}^{\mathbf{a}} \sum_{\mathbf{n} \in \mathcal{N}} x(\mathbf{Rn} - \mathbf{a})(\mathbf{z}^{\mathbf{R}})^{-\mathbf{n}}.$$

Let

$$X_{\mathbf{a}}^{(\mathbf{R})}(z) = \sum_{\mathbf{n} \in \mathcal{N}} x(\mathbf{Rn} - \mathbf{a})\mathbf{z}^{-\mathbf{n}} = \sum_{\mathbf{n} \in \mathcal{N}} x_{\mathbf{a}}(\mathbf{n})\mathbf{z}^{-\mathbf{n}},$$

then

$$X(z) = \sum_{\mathbf{a} \in \mathcal{N}(\mathbf{R})} \mathbf{z}^{\mathbf{a}} X_{\mathbf{a}}^{(\mathbf{R})}(\mathbf{z}^{\mathbf{R}}),$$

where $\mathbf{z}^{\mathbf{a}}$ advances the multidimensional signal by a vector amount $\mathbf{a}$. Thus, the set of all integer vectors can be partitioned into $J(\mathbf{R})$ equivalence classes using either Type-I or Type-II multidimensional polyphase decomposition. Moreover, the success of these decompositions rests on the following theorem.

Theorem 2.2.2.3. (Division Theorem for Integer Vectors) *Let* $\mathbf{R}$ *be a* $k \times k$ *nonsingular integer matrix, let* $\mathbf{p}$ *and* $\mathbf{n}$ *be an integer vectors, and let* $\mathbf{a}$ *be a shift vector of* $\mathbf{R}$. *Then, we can uniquely express* $\mathbf{p}$ *as*

$$\mathbf{p} = \mathbf{Rn} + \mathbf{a}.$$

Proof: Write $\mathbf{p}$ as

$$\mathbf{p} = \mathbf{p}_F + \mathbf{p}_L$$

where, $\mathbf{p}_F$ and $\mathbf{p}_L$ are unique vectors with

$$\mathbf{p}_F \in \text{FPD}(\mathbf{R}) \text{ and } \mathbf{p}_L \in \text{LAT}(\mathbf{R}).$$

Since $\mathbf{R}$ is an integer matrix, then $\mathbf{p}_L$ is an integer vector. Moreover, since $\mathbf{p}$ and $\mathbf{p}_L$ are integer vectors, then $\mathbf{p}_F$ is an integer vector. Moreover, since $\mathbf{p}_F \in \text{FPD}(\mathbf{R})$, it follows that $\mathbf{p}_F \in \mathcal{N}(\mathbf{R})$. By letting $\mathbf{p}_F = \mathbf{a}$ and $\mathbf{p}_L = \mathbf{Rn}$, we obtain

$$\mathbf{p} = \mathbf{Rn} + \mathbf{a}.$$

$\blacksquare$

The *remainder*, $\mathbf{a}$, can be expressed as $\mathbf{a} \equiv \mathbf{p} \mod \mathbf{R}$ or simply $((\mathbf{a})) = ((\mathbf{p}))_{\mathbf{R}}$.

Figure 2.8. Multidimensional expander.

2.3 Multidimensional building blocks

This section defines and analyzes two types of multidimensional building blocks used in multidimensional multirate signal processing. One type of building block deals with changes in the sampling rate of the input signal, and the other type of building block deals with changes in filter length. Analogous to the one-dimensional case, there are two types of multidimensional building blocks, which change the sampling rate of the input signal — decimators to reduce it and expanders to increase it. But in the multidimensional case, decimators and expanders affect not only the sampling rate but also the geometry of the sampling lattice. Then, a multidimensional building block is discussed which changes the length of a given filter — multidimensional comb filters to increase its length. In this way, a multidimensional comb filter version of a given filter is analogous to a multidimensional expander acting on an input signal.

2.3.1 Multidimensional expanders

Definition 2.3.1.1. *Let* $x(\mathbf{n})$ *be a* k-*dimensional signal and let* $\mathbf{L}$ *be an integer-valued matrix. Then, the multidimensional process of* $\mathbf{L}$-*fold expanding maps a signal on* $\mathcal{N}$ *to another signal that is nonzero only at points on the sublattice LAT(*$\mathbf{L}$*). The output of the* k-*dimensional expander is related to the input signal* $x(\mathbf{n})$ *by*

$$y(\mathbf{n}) = \begin{cases} x(\mathbf{L}^{-1}\mathbf{n}), & \textit{if } \mathbf{L}^{-1}\mathbf{n} \in \mathcal{N} \\ 0, & \textit{otherwise} \end{cases}$$

where $\mathbf{L}$ *is a nonsingular* $k \times k$ *integer matrix.*

Since LAT($\mathbf{L}$) denotes all vectors of the form $\mathbf{Lm}$, where $\mathbf{m} \in \mathcal{N}$, then the condition $\mathbf{L}^{-1}\mathbf{n} \in \mathcal{N}$ is equivalent to $\mathbf{n} \in$ LAT($\mathbf{L}$). The matrix $\mathbf{L}$ is known as the expansion matrix. The multidimensional expander is depicted pictorially in Figure 2.8. Now, let us analyze the multidimensional expander using the definition of the multidimensional z-transform.

$$Y(\mathbf{z}) = \sum_{\mathbf{n} \in \mathcal{N}} y(\mathbf{n})\mathbf{z}^{-\mathbf{n}}.$$

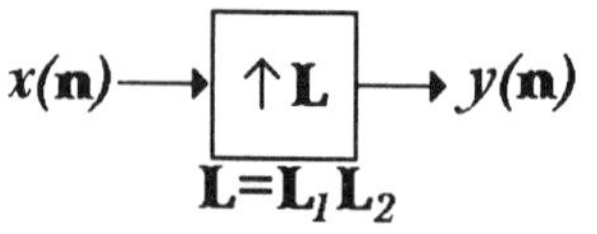

$$x(\mathbf{n}) \longrightarrow \boxed{\uparrow \mathbf{L}} \longrightarrow y(\mathbf{n})$$
$$\mathbf{L} = \mathbf{L}_l \mathbf{L}_2$$

Figure 2.9. Cascade of multidimensional expanders.

Then, applying the definition of the multidimensional expander will require that $Y(\mathbf{z})$ is zero for the lattice points defined by $\mathbf{L}^{-1}\mathbf{n} \notin \mathcal{N}$. So, let $\mathbf{n} = \mathbf{Lm}$. Then,

$$Y(\mathbf{z}) = \sum_{\mathbf{m} \in \mathcal{N}} y(\mathbf{Lm}) \mathbf{z}^{-\mathbf{Lm}}.$$

By the definition of the multidimensional expander, $x(\mathbf{m}) = y(\mathbf{Lm})$ for all $\mathbf{m} \in \mathcal{N}$. Hence,

$$Y(\mathbf{z}) = \sum_{\mathbf{m} \in \mathcal{N}} x(\mathbf{m}) \mathbf{z}^{-\mathbf{Lm}},$$

or equivalently, using Theorem 2.2.2.1, the equation becomes

$$Y(\mathbf{z}) = \sum_{\mathbf{m} \in \mathcal{N}} x(\mathbf{m}) \left(\mathbf{z}^{\mathbf{L}} \right)^{-\mathbf{m}}.$$

Hence,

$$Y(\mathbf{z}) = X(\mathbf{z}^{\mathbf{L}}).$$

In order to investigate the behavior of the multidimensional expander in the Fourier domain, replace $\mathbf{z}$ with $\exp(j\underline{\omega})$ to yield

$$Y\left(\exp(j\underline{\omega})\right) = X\left((\exp(j\underline{\omega}))^{\mathbf{L}} \right).$$

Utilizing Theorem 2.2.2.2 this equation becomes

$$Y\left(\exp(j\underline{\omega})\right) = X\left(\exp(j\mathbf{L}^{T}\underline{\omega}) \right).$$

By suppressing the exponential and, as such, changing the notation so that $Y(\underline{\omega})$ denotes $Y\left(\exp(j\underline{\omega})\right)$, then

$$Y(\underline{\omega}) = X(\mathbf{L}^{T}\underline{\omega}).$$

Now consider the cascade of two expanders $\mathbf{L}_1$ and $\mathbf{L}_2$, which can be depicted graphically by Figure 2.9. Substituting $\mathbf{L} = \mathbf{L}_1\mathbf{L}_2$ into the equation for $Y(\underline{\omega})$ yields

$$Y(\underline{\omega}) = X((\mathbf{L}_1\mathbf{L}_2)^{T}\underline{\omega}),$$

or simply,

$$Y(\underline{\omega}) = X(\mathbf{L}_2^{T}\mathbf{L}_1^{T}\underline{\omega}),$$

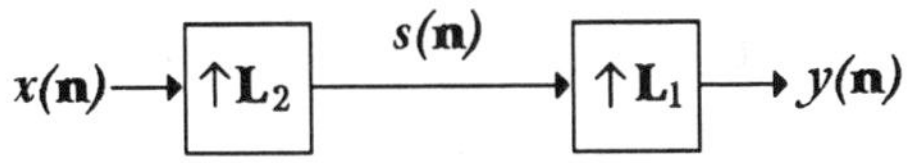

Figure 2.10. Interpreting cascaded multidimensional expanders.

$$x(\mathbf{n}) \rightarrow \boxed{\uparrow \mathbf{V_L}} \rightarrow \boxed{\uparrow \mathbf{\Lambda_L}} \rightarrow \boxed{\uparrow \mathbf{U_L}} \rightarrow y(\mathbf{n})$$

Figure 2.11. Smith form cascade of expanders.

which can be represented by Figure 2.10.

This cascade of two multidimensional expanders can be verified by re-alizing that

$$Y(\underline{\omega}) = S(\mathbf{L}_1^T \underline{\omega})$$

and

$$S(\underline{\omega}) = X(\mathbf{L}_2^T \underline{\omega}).$$

Therefore,

$$
\begin{aligned}
Y(\underline{\omega}) &= X(\mathbf{L}_2^T(\mathbf{L}_1^T \underline{\omega})) \\
&= X((\mathbf{L}_1 \mathbf{L}_2)^T \underline{\omega}) \\
&= X(\mathbf{L}^T \underline{\omega}).
\end{aligned}
$$

So, for example, if $\mathbf{L}$ is replaced by its Smith form decomposition, that is, $\mathbf{L} = \mathbf{U_L \Lambda_L V_L}$, then we can graphically depict it in Figure 2.11.

2.3.2 Multidimensional decimators

Definition 2.3.2.1. *Let $x(\mathbf{n})$ be a k-dimensional signal and let $\mathbf{M}$ be a k x k integer-valued matrix. The multidimensional $\mathbf{M}$-fold decimator samples input $x(\mathbf{n})$ by mapping points on the sublattice $LAT(\mathbf{M})$ to $\mathcal{N}$ according to*

$$y(\mathbf{n}) = x(\mathbf{Mn})$$

and discarding samples of $x(\mathbf{n})$ not on $LAT(\mathbf{M})$. $\mathbf{M}$ is called the decimation matrix.

The multidimensional decimator is depicted pictorially in Figure 2.12. Let us analyze the multidimensional decimator using the definition of the

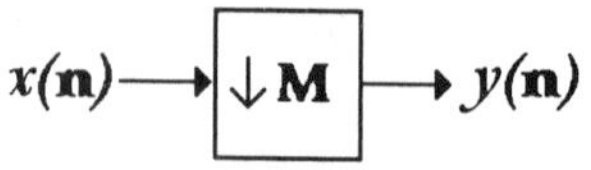

$$x(n) \longrightarrow \boxed{\downarrow \mathbf{M}} \longrightarrow y(n)$$

Figure 2.12. Multidimensional decimator.

multidimensional z-transform, *i.e.*

$$Y(\mathbf{z}) = \sum_{\mathbf{n} \in \mathcal{N}} y(\mathbf{n}) \mathbf{z}^{-\mathbf{n}}.$$

Substituting the definition of the multidimensional decimator, we obtain

$$Y(\mathbf{z}) = \sum_{\mathbf{n} \in \mathcal{N}} x(\mathbf{Mn}) \, \mathbf{z}^{-\mathbf{n}}.$$

Note that $x(\mathbf{n})$ is not zero for noninteger multiples of $\mathbf{Mn}$. So, define an intermediate mapping of points that is zero for noninteger multiples of $\mathbf{Mn}$, that is,

$$x_1(\mathbf{n}) = \begin{cases} x(\mathbf{n}), & \text{where } \mathbf{M}^{-1}\mathbf{n} \in \mathcal{N} \\ 0, & \text{otherwise .} \end{cases}$$

So, $y(\mathbf{n}) = x(\mathbf{Mn}) = x_1(\mathbf{Mn})$. Hence,

$$Y(\mathbf{z}) = \sum_{\mathbf{n} \in \mathcal{N}} x_1(\mathbf{Mn}) \, \mathbf{z}^{-\mathbf{n}}.$$

Let $\mathbf{m} = \mathbf{Mn}$. Then, by the definition of $x_1(\mathbf{m})$, if $\mathbf{M}^{-1}\mathbf{n} \notin \mathcal{N}$, then $x_1(\mathbf{m}) = 0$. Therefore,

$$Y(\mathbf{z}) = \sum_{\mathbf{m} \in \mathcal{N}} x_1(\mathbf{m}) \, \mathbf{z}^{-\mathbf{M}^{-1}\mathbf{m}},$$

or equivalently,

$$Y(\mathbf{z}) = \sum_{\mathbf{m} \in \mathcal{N}} x_1(\mathbf{m}) \, (\mathbf{z}^{\mathbf{M}^{-1}})^{-\mathbf{m}}.$$

Hence,

$$Y(\mathbf{z}) = X_1(\mathbf{z}^{\mathbf{M}^{-1}}).$$

Next, we need to express $X_1(\mathbf{z})$ in terms of $X(\mathbf{z})$. By the definition of $x_1(\mathbf{n})$, we can write the following:

$$x_1(\mathbf{n}) = C_{\mathbf{M}}(\mathbf{n}) x(\mathbf{n}),$$

where $C_{\mathbf{M}}(\mathbf{n})$ is a scalar-valued sampling function associated with the sampling matrix $\mathbf{M}$, that is,

$$C_{\mathbf{M}}(\mathbf{n}) = \begin{cases} 1, & \text{whenever } \mathbf{M}^{-1}\mathbf{n} \in \mathcal{N} \\ 0, & \text{otherwise} . \end{cases}$$

Since $C_{\mathbf{M}}(\mathbf{n})$ is periodic in $\text{LAT}(\mathbf{M})$ with spatial variable $\mathbf{n}$, *i.e.*

$$C_{\mathbf{M}}(\mathbf{n}) = C_{\mathbf{M}}(\mathbf{n} + \mathbf{Mm}),$$

it can be expressed as a complex Fourier series

$$C_{\mathbf{M}}(\mathbf{n}) = \frac{1}{J(\mathbf{M})} \sum_{\mathbf{h} \in \mathcal{I}_{\mathbf{M}^T}} \exp[j((2\pi\mathbf{M}^{-\mathbf{T}})\mathbf{h})^{\mathbf{T}}\mathbf{n}] .$$

Applying the definition of the multidimensional z-transform to $x_1(\mathbf{n})$ yields

$$X_1(\mathbf{z}) = \sum_{\mathbf{n} \in \mathcal{N}} x_1(\mathbf{n})\mathbf{z}^{-\mathbf{n}} = \sum_{\mathbf{n} \in \mathcal{N}} C_{\mathbf{M}}(\mathbf{n})x(\mathbf{n})\mathbf{z}^{-\mathbf{n}}.$$

Substituting the equation for $C_{\mathbf{M}}(\mathbf{n})$ yields

$$X_1(\mathbf{z}) = \frac{1}{J(\mathbf{M})} \sum_{\mathbf{h} \in \mathcal{I}_{\mathbf{M}^T}} \sum_{\mathbf{n} \in \mathcal{N}} \exp[j((2\pi\mathbf{M}^{-\mathbf{T}})\mathbf{h})^{\mathbf{T}}\mathbf{n}]x(\mathbf{n})\mathbf{z}^{-\mathbf{n}},$$

or equivalently,

$$X_1(\mathbf{z}) = \frac{1}{J(\mathbf{M})} \sum_{\mathbf{h} \in \mathcal{I}_{\mathbf{M}^T}} \left[\sum_{\mathbf{n} \in \mathcal{N}} x(\mathbf{n}) \left(\mathbf{z} \exp[-j(2\pi\mathbf{M}^{-\mathbf{T}})\mathbf{h}] \right)^{-\mathbf{n}} \right] .$$

Then, performing a multidimensional z-transform yields

$$X_1(\mathbf{z}) = \frac{1}{J(\mathbf{M})} \sum_{\mathbf{h} \in \mathcal{I}_{\mathbf{M}^T}} X(\mathbf{z} \exp[-j(2\pi\mathbf{M}^{-\mathbf{T}})\mathbf{h}]).$$

Since $Y(\mathbf{z}) = X_1(\mathbf{z}^{\mathbf{M}^{-1}})$, then $Y(\mathbf{z})$ becomes

$$Y(\mathbf{z}) = \frac{1}{J(\mathbf{M})} \sum_{\mathbf{h} \in \mathcal{I}_{\mathbf{M}^T}} X(\mathbf{z}^{\mathbf{M}^{-1}} \exp[-j(2\pi\mathbf{M}^{-\mathbf{T}})\mathbf{h}]).$$

If $\mathbf{z}^{-\mathbf{n}}$ equals $\exp(-j\underline{\omega}^T\mathbf{n})$, then $\mathbf{z}^{\mathbf{M}^{-1}}$ is equivalent to $\exp(-j\mathbf{M}^{-T}\underline{\omega})$. Therefore,

$$Y(\exp[j\underline{\omega}]) = \frac{1}{J(\mathbf{M})} \sum_{\mathbf{h} \in \mathcal{I}_{\mathbf{M}^T}} X((\exp[j\underline{\omega}])^{\mathbf{M}^{-1}} \exp[-j2\pi\mathbf{M}^{-T}\mathbf{h}]).$$

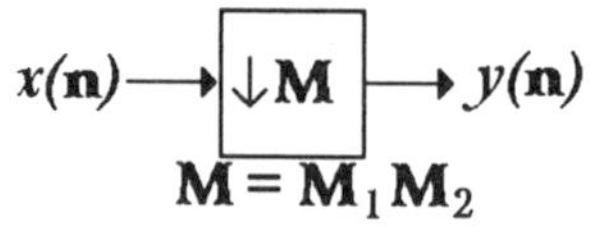

$$x(n) \longrightarrow \boxed{\downarrow \mathbf{M}} \longrightarrow y(n)$$
$$\mathbf{M} = \mathbf{M}_1 \mathbf{M}_2$$

Figure 2.13. Cascade of multidimensional decimators.

$$x(\mathbf{n}) \longrightarrow \boxed{\downarrow \mathbf{M}_1} \xrightarrow{\quad s(n) \quad} \boxed{\downarrow \mathbf{M}_2} \longrightarrow y(\mathbf{n})$$

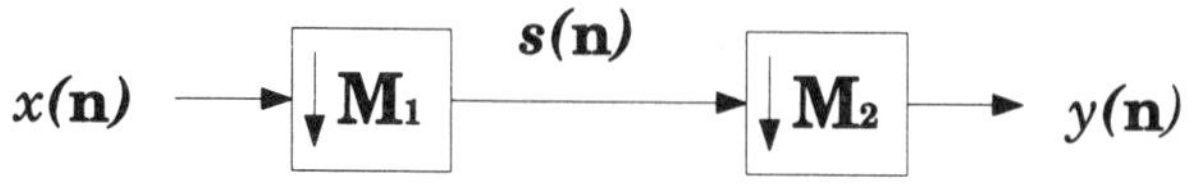

Figure 2.14. Interpreting the cascade of multidimensional decimators.

Utilizing Theorem 2.2.2.2, $Y(\exp[j\,\underline{\omega}])$ becomes

$$Y(\exp[j\,\underline{\omega}]) = \frac{1}{J(\mathbf{M})} \sum_{\mathbf{h} \in \mathcal{I}_{\mathbf{M}^T}} X(\exp[j\mathbf{M}^{-T}(\underline{\omega} - 2\pi\mathbf{h})]).$$

By suppressing the exponential and, as such, changing the notation so that $Y(\underline{\omega})$ denotes $Y(\exp[j\,\underline{\omega}])$, then

$$Y(\underline{\omega}) = \frac{1}{J(\mathbf{M})} \sum_{\mathbf{h} \in \mathcal{I}_{\mathbf{M}^T}} X[\mathbf{M}^{-T}(\underline{\omega} - 2\pi\mathbf{h})].$$

Now consider the cascade of two decimators $\mathbf{M}_1$ and $\mathbf{M}_2$, which can be depicted graphically in Figure 2.13. Substituting $\mathbf{M} = \mathbf{M}_1\mathbf{M}_2$ into the definition of the multidimensional decimator yields

$$y(\mathbf{n}) = x(\mathbf{M}_1\mathbf{M}_2\mathbf{n})$$

which can be represented in Figure 2.14. This cascade of multidimensional decimators can be verified by realizing that

$$y(\mathbf{n}) = s(\mathbf{M}_2\mathbf{n})$$

and

$$s(\mathbf{n}) = x(\mathbf{M}_1\mathbf{n}).$$

Therefore,

$$y(\mathbf{n}) = x(\mathbf{M}_1\mathbf{M}_2\mathbf{n}).$$

So, for example, if $\mathbf{M}$ is replaced by its Smith form decomposition, that is, $\mathbf{M} = \mathbf{U}_\mathbf{M}\mathbf{\Lambda}_\mathbf{M}\mathbf{V}_\mathbf{M}$, then we can graphically depict it as Figure 2.15.

Let us briefly examine unimodular decimators with decimation matrix $\mathbf{V}$. Consider multidimensional decimator,

Figure 2.15. Smith form cascade of decimators.

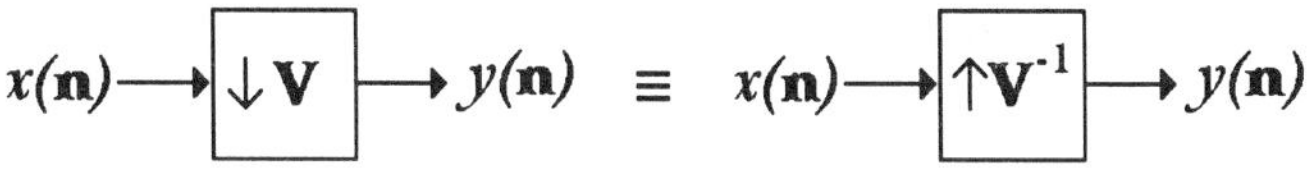

$$x(\mathbf{n}) \longrightarrow \boxed{\downarrow \mathbf{V}} \longrightarrow y(\mathbf{n}) \quad \equiv \quad x(\mathbf{n}) \longrightarrow \boxed{\uparrow \mathbf{V}^{-1}} \longrightarrow y(\mathbf{n})$$

Figure 2.16. Unimodular decimator.

$$Y(\mathbf{z}) = \frac{1}{J(\mathbf{V})} \sum_{\mathbf{h} \in \mathcal{I}_{\mathbf{V}^T}} X(\mathbf{z}^{\mathbf{V}^{-1}} \exp[-j(2\pi \mathbf{V}^{-T})\mathbf{h}]).$$

Since $\mathbf{V}$ is a unimodular, then $J(\mathbf{V}) = 1$. Therefore,

$$Y(\mathbf{z}) = X(\mathbf{z}^{\mathbf{V}^{-1}}).$$

Therefore, there is no aliasing and, as such, unimodular decimation can be viewed as just a rearrangement of samples. This interpretation of unimodular decimation is depicted in Figure 2.16.

2.3.3 Multidimensional comb filters

Definition 2.3.3.1. *Given the impulse response $h(\mathbf{n})$ of a multidimensional filter, one can build a multidimensional comb filter $g(\mathbf{n})$ by mapping the filter $h(\mathbf{n})$ on $\mathcal{N}$ to another one that is nonzero only at points on the sublattice $LAT(\mathbf{L})$, i.e.*

$$g(\mathbf{n}) = \begin{cases} h(\mathbf{L}^{-1}\mathbf{n}), & \text{if } \mathbf{L}^{-1}\mathbf{n} \in \mathcal{N} \\ 0, & \text{otherwise}. \end{cases}$$

As mentioned earlier, $\mathbf{L}^{-1}\mathbf{n} \in \mathcal{N}$ is equivalent to $\mathbf{n} \in LAT(\mathbf{L})$. The impulse response of a multidimensional comb filter can be represented by

$$g(\mathbf{n}) = \sum_{\mathbf{m} \in \mathcal{N}} h(\mathbf{m}_i) \delta_{\mathbf{n}, \mathbf{Lm}}$$

where $\delta_{\mathbf{n}, \mathbf{Lm}}$ is the multidimensional Kronecker delta function. Taking the multidimensional z-transform of $g(\mathbf{n})$, we obtain

$$G(\mathbf{z}) = \sum_{\mathbf{n} \in \mathcal{N}} \sum_{\mathbf{m} \in \mathcal{N}} h(\mathbf{m}) \delta_{\mathbf{n}, \mathbf{Lm}} \mathbf{z}^{-\mathbf{n}},$$

or equivalently,

$$G(\mathbf{z}) = \sum_{\mathbf{m} \in \mathcal{N}} h(\mathbf{m}) \mathbf{z}^{-\mathbf{Lm}}$$

or simply,

$$G(\mathbf{z}) = H(\mathbf{z}^{\mathbf{L}}).$$

If LAT(**L**) is not rectangular, that is, **L** is not a diagonal matrix, then the components of $H(\mathbf{z}^{\mathbf{L}})$ are no longer separable. To illustrate this point, let $\mathbf{L} = \begin{bmatrix} x_0 & x_1 \\ y_0 & y_1 \end{bmatrix}$, then $H(\mathbf{z}^{\mathbf{L}}) = H(z_0^{x_0} z_1^{y_0}, z_0^{x_1} z_1^{y_1})$ and $H(\mathbf{z}) = H(z_0, z_1)$. For example, we will consider both rectangular sampling and hexagonal sampling. For rectangular sampling, let

$$\mathbf{L} = \begin{bmatrix} 2 & 0 \\ 0 & 3 \end{bmatrix},$$

then $H(\mathbf{z}^{\mathbf{L}}) = H(z_0^2 z_1^0, z_0^0 z_1^3) = H(z_0^2, z_1^3)$ and $H(\mathbf{z}) = H(z_0, z_1)$. This indicates the separability of components of $H(\mathbf{z}^{\mathbf{L}})$ for a rectangular sampling. For hexagonal sampling, let

$$\mathbf{L} = \begin{bmatrix} 1 & 1 \\ 2 & -2 \end{bmatrix},$$

then $H(\mathbf{z}^{\mathbf{L}}) = X(z_0^1 z_1^2, z_0^1 z_1^{-2}) = H(z_0 z_1^2, z_0 z_1^{-2})$ and $H(\mathbf{z}) = H(z_0, z_1)$. This indicates the lack of separability of components of the multidimensional comb filter $H(\mathbf{z}^{\mathbf{L}})$ for non-rectangular sampling.

2.4 Interchanging building blocks

This section defines ways to interchange multidimensional building blocks. First, the conditions for interchanging multidimensional decimators and expanders for sampling rate conversion are presented. Secondly, we consider the multidimensional noble identities, an approach for interchanging multidimensional building blocks with multidimensional decimation and interpolation filters.

2.4.1 Interchanging decimators and expanders

Multidimensional sampling rate conversion is important for many signal processing applications, because many times it is necessary to interface image or video data between systems which use different sampling lattices. Examples include the conversion between European and American television systems and the conversion between high definition television (HDTV)

Figure 2.17. Decimator preceding expander.

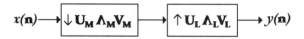

Figure 2.18. Substituting Smith form decompositions.

signals and conventional television signals. Thus, sampling rate conversion will require the cascade of a multidimensional L-fold expander and a multidimensional M-fold decimator, separated by a filter. Under what conditions can the decimator and the expander be interchanged? We will see that we will need to assume that (a) $\mathbf{M}$ and $\mathbf{L}$ commute, that is, $\mathbf{LM} = \mathbf{ML}$, and (b) $\mathbf{M}$ and $\mathbf{L}$ are coprime. We will see that through the utilization of the Smith form that coprimeness in multidimensions can be achieved in each dimension independently.

Consider the configuration in which the decimator precedes the expander, which is depicted in Figure 2.17. Let us assume the following Smith form decompositions for $\mathbf{M}$ and $\mathbf{L}$:

$$\mathbf{M} = \mathbf{U_M}\mathbf{\Lambda_M}\mathbf{V_M} \text{ and } \mathbf{L} = \mathbf{U_L}\mathbf{\Lambda_L}\mathbf{V_L}$$

where $\mathbf{U_M}, \mathbf{V_M}, \mathbf{U_L}, \mathbf{V_L}$ are unimodular matrices and $\mathbf{\Lambda_M}, \mathbf{\Lambda_L}$ are diagonal matrices. Therefore, the resulting configuration is depicted in Figure 2.18, or equivalently in Figure 2.19.

Without loss of generality, we can assume that

$$\mathbf{V_M} = \mathbf{V_L}.$$

Hence, we obtain the configuration depicted in Figure 2.20. Since $\mathbf{\Lambda_M}$ and

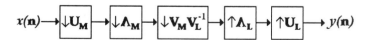

Figure 2.19. Substituting cascade definitions.

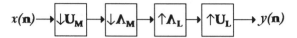

Figure 2.20. Simplifying the cascade.

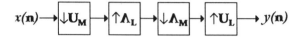

Figure 2.21. Interchanging lambda matrices.

Λ_L are diagonal matrices, then interchanging Λ_M and Λ_L can be achieved in each dimension individually, provided that the associated decimation and expansion ratios in each dimension are relatively prime. In this way, the Smith form decomposition has helped us transform a multidimensional problem into a series of one-dimensional problems. Therefore, we obtain the configuration depicted in Figure 2.21. Since unimodular decimators are equivalent to inverse unimodular expanders (as depicted in Figure 2.22), then for any unimodular matrix $\mathbf{T}$, Figure 2.23 results. Now choose a unimodular matrix $\mathbf{U}_A$ such that

$$\mathbf{L} = \mathbf{U}_A \Lambda_L \mathbf{U}_M^{-1},$$

or equivalently,

$$\mathbf{U}_A = \mathbf{L} \mathbf{U}_M \Lambda_L^{-1}.$$

Choose a matrix $\mathbf{V}_B$ such that

$$\mathbf{M} = \mathbf{U}_A \Lambda_M \mathbf{U}_L^{-1} \mathbf{V}_B,$$

or equivalently,

$$\mathbf{V}_B = \mathbf{U}_L \Lambda_M^{-1} \mathbf{U}_A^{-1} \mathbf{M}.$$

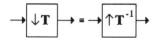

Figure 2.22. Unimodular decimators and unimodular expanders.

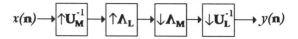

Figure 2.23. Resulting figure.

Substituting U_A^{-1} into the last equation yields

$$V_B = U_L \Lambda_M^{-1} \Lambda_L U_M^{-1} L^{-1} M.$$

Since Λ_M and Λ_L are relatively prime in each dimension, we can interchange Λ_M^{-1} and Λ_L. Therefore,

$$V_B = U_L \Lambda_L \Lambda_M^{-1} U_M^{-1} L^{-1} M.$$

Since $V_M = V_L$, then $V_L V_M^{-1} = I$. Therefore,

$$V_B = U_L \Lambda_L V_L V_M^{-1} \Lambda_M^{-1} U_M^{-1} L^{-1} M,$$

or equivalently,

$$V_B = L M^{-1} L^{-1} M.$$

If L and M are assumed to commute, then $L^{-1} M = M L^{-1}$. Therefore,

$$V_B = I.$$

Hence, we have shown that the multidimensional L-fold expander and the multidimensional M-fold decimator can commute provided (a) L and M commute and (b) L and M are coprime.

Example 2.4.1.1. Let us consider the following example. Assume a multidimensional decimator is defined by the decimation matrix D is given by

$$D = \begin{bmatrix} 1 & -1 \\ 1 & 2 \end{bmatrix}$$

and a multidimensional expander is defined by the expansion matrix E is given by

$$E = \begin{bmatrix} 2 & 0 \\ 0 & 2 \end{bmatrix}.$$

Can these multidimensional building blocks be interchanged? First, we will find the Smith form decomposition of matrices D and E, that is,

$$D = U_D \Lambda_D V_D = \begin{bmatrix} 1 & 0 \\ 1 & 1 \end{bmatrix} \begin{bmatrix} 1 & 0 \\ 0 & 3 \end{bmatrix} \begin{bmatrix} 1 & -1 \\ 0 & 1 \end{bmatrix}$$

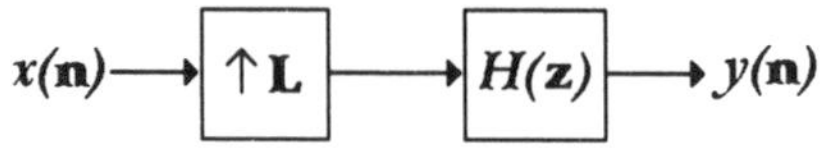

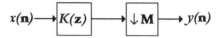

Figure 2.24. Multidimensional expander followed by filter.

$$x(\mathbf{n}) \longrightarrow \boxed{K(\mathbf{z})} \longrightarrow \boxed{\downarrow \mathbf{M}} \longrightarrow y(\mathbf{n})$$

Figure 2.25. Multidimensional decimator preceded by filter.

and

$$E = U_E \Lambda_E V_E = \begin{bmatrix} 1 & 1 \\ 0 & 1 \end{bmatrix} \begin{bmatrix} 2 & 0 \\ 0 & 2 \end{bmatrix} \begin{bmatrix} 1 & -1 \\ 0 & 1 \end{bmatrix}.$$

Since the diagonal entries in Λ_D and Λ_E are relatively prime, then Λ_D and Λ_E are coprime, which in turn implies that D and E are coprime. Secondly, we need to check to see whether D and E commute, that is,

$$DE = \begin{bmatrix} 1 & -1 \\ 1 & 2 \end{bmatrix} \begin{bmatrix} 2 & 0 \\ 0 & 2 \end{bmatrix} = \begin{bmatrix} 2 & -2 \\ 2 & 4 \end{bmatrix}$$

and

$$ED = \begin{bmatrix} 2 & 0 \\ 0 & 2 \end{bmatrix} \begin{bmatrix} 1 & -1 \\ 1 & 2 \end{bmatrix} = \begin{bmatrix} 2 & -2 \\ 2 & 4 \end{bmatrix}.$$

Since $DE = ED$, then D and E commute. Since D and E are coprime and since D and E commute, then the multidimensional decimator (defined by the decimation matrix D) and the multidimensional expander (defined by the expansion matrix E) can be interchanged.

2.4.2   Multidimensional noble identities

In most applications involving multidimensional interpolation, an interpolation filter follows an expander as in Figure 2.24. Similarly, in many applications involving multidimensional decimators, a decimation filter precedes the decimator as in Figure 2.25. If $H(\mathbf{z})$ and $K(\mathbf{z})$ are multidimensional comb filters, then can the multidimensional building blocks be interchanged?

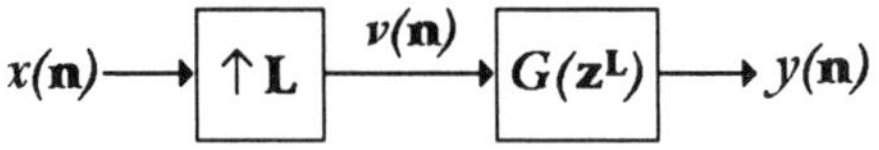

$$x(n)\longrightarrow \boxed{\uparrow L}\xrightarrow{v(n)}\boxed{G(z^L)}\longrightarrow y(n)$$

Figure 2.26. Multidimensional expander with comb filter.

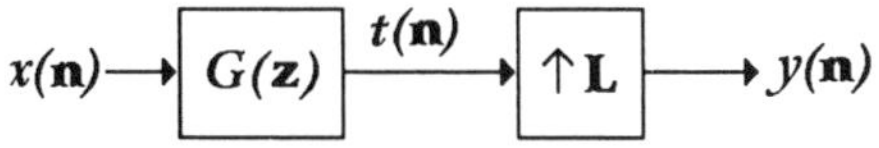

$$x(n)\longrightarrow \boxed{G(z)}\xrightarrow{t(n)}\boxed{\uparrow L}\longrightarrow y(n)$$

Figure 2.27. Filter preceding multidimensional expander.

2.4.2.1 Interchanging filters and expanders

Consider the configuration depicted in Figure 2.26. Writing the elemental equations, we obtain

$$V(z) = X(z^L)$$

and

$$Y(z) = G(z^L)V(z).$$

Hence,

$$Y(z) = G(z^L)X(z^L).$$

But, this equation could also be interpreted as Figure 2.27, where

$$Y(z) = T(z^L)$$

and

$$T(z) = G(z)X(z).$$

These two equivalent block diagrams present a systematic approach for interchanging multidimensional filters with multidimensional expanders, and together they will be referred to as the Noble identity for multidimensional expanders.

2.4.2.2 Interchanging filters and decimators

Consider the configuration depicted in Figure 2.28. Writing the elemental equations,

$$V(z) = G(z^M)\,X(z)$$

and

$$Y(z) = \frac{1}{J(M)}\sum_{h\in\mathcal{I}_{M^T}} V(z^{M^{-1}}\exp[-j(2\pi M^{-T})h]).$$

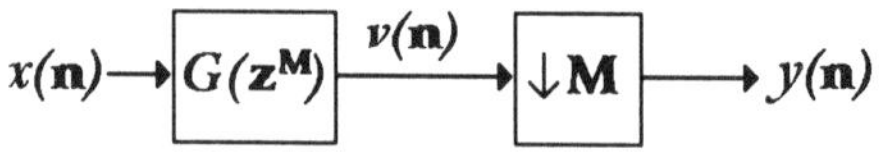

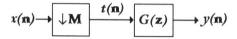

Figure 2.28. Comb filter preceding multidimensional decimator.

$$x(\mathbf{n}) \rightarrow \boxed{\downarrow \mathbf{M}} \xrightarrow{t(\mathbf{n})} \boxed{G(\mathbf{z})} \rightarrow y(\mathbf{n})$$

Figure 2.29. Multidimensional decimator preceding filter.

Hence, using the multidimensional z-transform identities associated with $\mathbf{z}$ raised to a matrix power, we obtain the following

$$Y(\mathbf{z}) = \frac{G(\mathbf{z})}{J(\mathbf{M})} \sum_{\mathbf{h} \in \mathcal{I}_{\mathbf{M}^T}} X(\mathbf{z}^{\mathbf{M}^{-1}} \exp[-j(2\pi \mathbf{M}^{-T})\mathbf{h}]).$$

But this equation could be interpreted as Figure 2.29, where

$$Y(\mathbf{z}) = G(\mathbf{z})T(\mathbf{z})$$

and

$$T(\mathbf{z}) = \frac{1}{J(\mathbf{M})} \sum_{\mathbf{h} \in \mathcal{I}_{\mathbf{M}^T}} X(\mathbf{z}^{\mathbf{M}^{-1}} \exp[-j(2\pi \mathbf{M}^{-T})\mathbf{h}]).$$

These two equivalent block diagrams present a systematic approach for interchanging multidimensional filters with multidimensional decimators, and together they will be referred to as the Noble identity for multidimensional decimators.

2.5   Problems

1. Consider the following sampling matrix:

$$\mathbf{T} = \begin{bmatrix} 2 & 2 \\ 4 & -4 \end{bmatrix}.$$

Sketch the lattice LAT($\mathbf{T}$). Clearly indicate the fundamental parallelepiped FPD($\mathbf{T}$) and highlight the J($\mathbf{T}$) points in $\mathcal{N}$ which belong to FPD($\mathbf{T}$).

2. Consider the arbitrary sampling lattice given by

$$\mathbf{V} = \begin{bmatrix} a & b \\ 0 & c \end{bmatrix}.$$

What is the corresponding reciprocal lattice?

3. Consider the following matrix:

$$\mathbf{R} = \begin{bmatrix} 2 & -2 \\ 2 & 6 \end{bmatrix}.$$

Sketch the lattice generated by $\mathbf{R}$. What is $J(\mathbf{R})$? Sketch $FPD(\mathbf{R})$ and $FPD(\mathbf{R}^T)$. Specify the elements of $\mathcal{N}(\mathbf{R})$ and $\mathcal{N}(\mathbf{R}^T)$.

4. Consider the following matrix:

$$\mathbf{R} = \begin{bmatrix} 1 & 1 \\ -1 & 1 \end{bmatrix}.$$

Using the elementary row and column operations, find the Smith form decomposition of $\mathbf{R}$.

5. Show that the scaling property for k-dimensional z-transforms is given by the following: If

$$y(\mathbf{n}) = \mathbf{a}^{-\mathbf{n}} x(\mathbf{n})$$

then

$$Y(\mathbf{z}) = X(\mathbf{A}\mathbf{z}),$$

where,

$$\mathbf{a} = [a_1, \ldots, a_k]^T$$

and

$$\mathbf{A} = \operatorname{diag}(a_1, \ldots, a_k).$$

6. If a k-dimensional signal $x(\mathbf{n})$ is separable, that is,

$$x(\mathbf{n}) = x_1(n_1), x_2(n_2), \ldots, x_k(n_k),$$

then what, if anything, can be said about the separability of the DFT?

7. Let us generalize the notion of multidimensional expanders and decimators. Let **N** be a nonsingular matrix. In the definition of the multidimensional expander, replace $\mathcal{N}$ with LAT(**N**), where LAT(**L**) $\subseteq$ LAT(**N**). Similarly, in the definition of the multidimensional decimator, replace $\mathcal{N}$ with LAT (**N**), where LAT(**M**) $\subseteq$ LAT(**N**).

   Interpret the resulting operators and provide examples to illustrate the use of each of them.

Chapter 3

Multirate Filter Banks

3.1 Introduction

This chapter introduces the notion of multirate filter banks. Using conventional filter design techniques, a high-order filter is required to obtain a fast roll-off capability. Filter banks are based on an alternative approach to realizing a high-order filter consisting of the cascade of lower-order (analysis) filters, that are designed with aliasing, and (synthesis) filters, that are designed to cancel the alias-components of the analysis filters.

Working independently, Smith and Barnwell[44] and Mintzer[35] reported the existence of a two-channel filter bank, which permitted perfect reconstruction of the input signal. Subsequently, Smith and Barnwell[45] developed the alias-component matrix formulation for analyzing M-channel filter banks. Meanwhile, Vetterli[53] and Vaidyanathan[48] independently discovered the polyphase formulation of analyzing M-channel filter banks. The theoretical underpinnings of filter banks were carefully examined by Vaidyanathan and Mitra[50], who showed the connection between pseudocirculant matrices and alias-free filter banks, and by Vaidyanathan[48], who showed a connection between paraunitary matrices and perfect reconstruction. These developments in multirate signal processing were recently complemented by research involving the joint consideration of quantization effects and filter bank design(Kovačević[25]; Westerink *et al.*[56]). Subsequently, Koilpillai and Vaidyanathan[24] developed perfect reconstructing cosine-modulated filter banks, which have the advantage that all filters are derived from a single prototype filter. In an effort to eliminate blocking effects in low-bit rate image compression, Malvar[29, 30] independently developed cosine modulated filter banks and he called them *lapped orthogonal transforms* (LOTs). Subsequently, using the very general conditions for windows, which were developed by Coifman and Meyer[9], Suter and Oxley[47], and independently Auscher, Weiss, and Wickerhauser[1] developed the continuous generalization of LOTs to permit the construction of different local bases in different time intervals. Yves Meyer called the continuous generalization of LOTs by the name Malvar wavelet[33], since they complement classical wavelet theory. Recently, Xia and Suter have generalized the theory of Malvar wavelets to include two-dimensional nonseparable Malvar wavelets[58] and Malvar wavelets on hexagons[59].

Vetterli[52] was the first person to write a paper on multidimensional multirate filter banks. His paper dealt with two-channel filter banks with quincunx decimation. Using an analogy with one dimensional systems, Viscito and Allebach[55] formalized the theory of multidimensional multirate operations for arbitrary multidimensional lattices. Karlsson and Vetterli[23] and Chen and Vaidyanathan[6] formulated polyphase representations of multidimensional signals. Some of the concepts developed in this chapter are also discussed in the texts by Fliege[19], by Strang and Nguyen[46], and by Vaidyanathan[49].

Section 3.2 introduces quadrature mirror filter banks. Then, Section 3.3 presents the theoretical foundations of multirate filter banks. Section 3.4 presents filter banks for spectral analysis. Section 3.5 introduces multidimensional quadrature mirror filter banks.

3.2   Quadrature mirror filter banks

This section defines and analyzes quadrature mirror filter (QMF) banks. The role of QMF banks in source coding is examined first. This is followed by a discussion of two important QMF bank formulations — aliascomponent and polyphase. Then, a multirate source coding design example is presented that illustrates many of the QMF bank concepts of this section. Finally, a brief discussion of quantization and their effects on filter banks is presented.

3.2.1   Source coding and QMF banks

Consider the two-channel QMF bank that is depicted in Figure 3.1. The analysis bank together with the decimators decompose the input signal $x(n)$ into two subband signals, $y_0(n)$ and $y_1(n)$. This is followed by expanders and the synthesis bank, which produce an output signal that reconstructs the input signal. For the QMF bank, the number of samples per unit time for the sum of both subband signals equals the number of samples per unit time for the input signal. But, the power of the subband signals is usually much lower than the original signal. After decoding, the input signal is reconstructed. Before we can design a system based on these ideas, we will need to become acquainted with the approaches that are used to analyze filter banks.

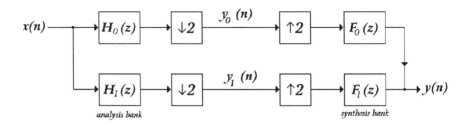

Figure 3.1. Two-channel filter bank.

3.2.2 Filter bank formulations

We will discuss two different approaches to the analysis of filter banks: (1) alias-component formulation, and (2) polyphase formulation. Then, we will show that the two formulations are equivalent.

3.2.2.1 Alias-component formulation of filter banks

Consider two-channel filter banks. Let $H_0(\omega)$ and $H_1(\omega)$ represent the frequency response of $H_0(z)$ and $H_1(z)$, respectively. For a QMF bank, the magnitude responses, $| H_0(\omega) |$ and $| H_1(\omega) |$, are mirror-images of each other with respect to frequency $\frac{\pi}{2}$,which is one quarter of the sampling frequency 2π. For M-channels ($M > 2$), this structure should not be called a QMF bank, because the traditional two-channel meaning does not hold. However, QMF has been used by many other authors, so we will retain the same nomenclature.

An M-channel QMF bank, that is depicted in Figure 3.2, partitions the signal spectra into M bands of equal bandwidth and later recombines these frequency bands. The input is denoted $x(n)$ and the output is denoted $y(n)$. It consists of M decimators (each with a decimation ratio of M); M expanders (each with an expansion ratio of M); M analysis filters (denoted by $\mathrm{H}_k(z)$, $k = 0, \ldots, M - 1$); and M synthesis filters (denoted $\mathrm{F}_k(z)$, $k = 0, \ldots, M - 1$).

The basic philosophy behind the design of QMF banks is to permit aliasing in the filters of the analysis bank and then choose the filters of the synthesis bank so that the alias-components in the filters of the analysis bank are cancelled.

Now, we will proceed with the analysis of the quadrature mirror filter bank, by examining it stage-by-stage. First, we will consider the analysis bank, which is depicted in Figure 3.3.

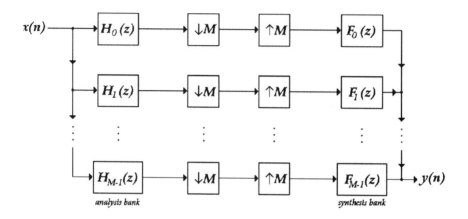

Figure 3.2. M-Channel filter bank.

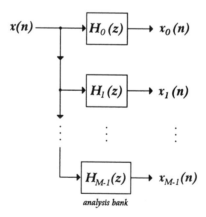

Figure 3.3. Analysis bank.

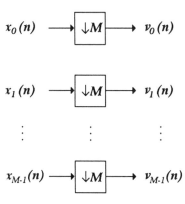

Figure 3.4. Bank of decimators.

The elemental equation for this stage is given by

$$X_k(z) = H_k(z)X(z)$$

for each $k = 0, 1, \ldots M - 1$. Then, we will consider the bank of decimators, which is depicted in Figure 3.4. The elemental equation for this stage is given by

$$V_k(z) = \frac{1}{M} \sum_{n=0}^{M-1} X_k(z^{1/M} W_M^n).$$

The following stage consists of a bank of expanders, which is depicted in Figure 3.5. The elemental equation for this stage is given by

$$U_k(z) = V_k(z^M).$$

Lastly, we will consider, the synthesis bank, which is depicted in Figure 3.6. The elemental equation for this stage is given by

$$Y(z) = \sum_{k=0}^{M-1} F_k(z)U_k(z).$$

Combining the equation for $X_k(z)$ with the equation for $V_k(z)$ yields

$$V_k(z) = \frac{1}{M} \sum_{n=0}^{M-1} H_k(z^{1/M} W_M^n)X(z^{1/M} W_M^n).$$

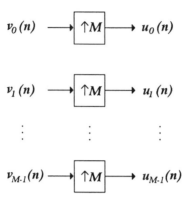

Figure 3.5. Bank of expanders.

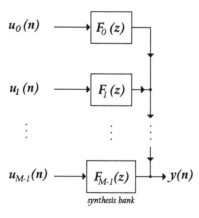

Figure 3.6. Synthesis bank.

Combining this equation with the equation for $U_k(z)$ yields

$$U_k(z) = \frac{1}{M} \sum_{n=0}^{M-1} H_k(zW_M^n)X(zW_M^n).$$

Finally, combining this equation with the equation for $Y(z)$ yields $Y(z) =$

$$\underbrace{\left[\frac{1}{M} \sum_{k=0}^{M-1} F_k(z)H_k(z)\right] X(z)}_{\text{desired terms}} + \underbrace{\sum_{n=1}^{M-1} \left[\frac{1}{M} \sum_{k=0}^{M-1} F_k(z)H_k(zW_M^n)\right] X(zW_M^n).}_{\text{terms due to aliasing}}$$

A few observations can be made. First, the desired term can be interpreted as the input signal weighted by the mean of the product of the analysis and synthesis filters. Secondly, if the coefficients of the aliasing term can be set to zero then a linear time invariant (LTI) system can be constructed out of linear time varying (LTV) components — decimators and expanders. In this case, when aliasing is cancelled the distortion function is given by

$$T(z) = \frac{1}{M} \sum_{k=0}^{M-1} F_k(z)H_k(z).$$

Unless $T(z)$ is an allpass filter, that is, $\mid T(\omega) \mid = c \neq 0$ for all ω, we say $Y(z)$ suffers from *amplitude distortion*. Similarly, unless $T(z)$ has linear phase, that is, $\arg T(\omega) = a + b\omega$ for constants a and b, we say $Y(z)$ suffers from *phase distortion*. This leads us to the definition of a perfect reconstruction system.

Definition 3.2.2.1. *Let $x(n)$ and $y(n)$ be the input and output, respectively, to the filter bank. Then, a perfect reconstruction system is a system free from aliasing, amplitude distortion, and phase distortion. As such, $y(n)$ is a scaled and delayed version of $x(n)$, that is,*

$$y(n) = cx(n - n_0).$$

Let us reexamine the output of the filter bank $Y(z)$, which was given by

$$Y(z) = \sum_{n=0}^{M-1} \left[\frac{1}{M} \sum_{k=0}^{M-1} F_k(z)H_k(zW_M^n)\right] X(zW_M^n),$$

or equivalently,

$$Y(z) = \sum_{n=0}^{M-1} A_n(z)X(zW_M^n)$$

where

$$A_n(z) = \frac{1}{M} \sum_{k=0}^{M-1} F_k(z) H_k(z W_M^n).$$

Writing the system of equations $A_n(z)$, $n = 0, \ldots, M - 1$, in matrix form yields

$$\mathbf{A}(z) = \frac{1}{M} \mathbf{H}(z) \mathbf{f}(z),$$

where,

$$\mathbf{A}(z) = [A_0(z),\ A_1(z), \ldots, A_{M-1}(z)]^T,$$

the synthesis bank is given by

$$\mathbf{f}(z) = [F_0(z),\ F_1(z), \ldots, F_{M-1}(z)]^T,$$

and the *so-called* alias-component (AC) matrix $\mathbf{H}(z)$ is given by

$$\mathbf{H}(z) = \begin{bmatrix} H_0(z) & H_1(z) & \cdots & H_{M-1}(z) \\ H_0(z W_M) & H_1(z W_M) & \cdots & H_{M-1}(z W_M) \\ \vdots & \vdots & \ddots & \vdots \\ H_0(z W_M^{M-1}) & H_1(z W_M^{M-1}) & \cdots & H_{M-1}(z W_M^{M-1}) \end{bmatrix}.$$

In this formulation, aliasing can be eliminated if and only if the gain for each of the aliasing terms equals zero, that is, $A_k(z) = 0$ for $k = 1, \ldots, M - 1$. Moreover, to insure *perfect reconstruction* $A_0(z)$ must be a delay, that is, $A_0(z) = z^{-m_0}$.

The solution of this system of equations for synthesis filters $\mathbf{f}(z)$, may have practical difficulties. It requires the inversion of the alias-component matrix $\mathbf{H}(z)$. Even if successful, that is, if $\mathbf{H}(z)$ is nonsingular, there is no guarantee that the resulting filters $\mathbf{f}(z)$ are stable, that is, the poles of $\mathbf{f}(z)$ are inside the unit circle. The approach given in the next section presents a different technique in which all of the above difficulties are removed.

Let us provide an example to illustrate the use of the alias-component matrix. Assume the analysis bank is characterized by the following filters:

$$H_0(z) = 1 \quad \text{and} \quad H_1(z) = z^{-1}.$$

Find the analysis filters $F_0(z)$ and $F_1(z)$ that satisfy the alias cancellation property. So, the alias-component formulation of the filter bank is characterized by

$$\mathbf{A}(z) = \frac{1}{M} \mathbf{H}(z) \mathbf{f}(z).$$

Writing this matrix equation in terms of its terms yields

$$\left[\begin{array}{c} A_0(z) \\ A_1(z) \end{array}\right] = \frac{1}{2} \left[\begin{array}{cc} H_0(z) & H_1(z) \\ H_0(zW_2) & H_1(zW_2) \end{array}\right] \left[\begin{array}{c} F_0(z) \\ F_1(z) \end{array}\right],$$

or equivalently,

$$\left[\begin{array}{c} A_0(z) \\ A_1(z) \end{array}\right] = \frac{1}{2} \left[\begin{array}{cc} H_0(z) & H_1(z) \\ H_0(-z) & H_1(-z) \end{array}\right] \left[\begin{array}{c} F_0(z) \\ F_1(z) \end{array}\right].$$

To insure alias cancellation, $A_0(z) = a(z)$ and $A_1(z) = 0$. Hence,

$$\mathbf{A}(z) = \left[\begin{array}{c} a(z) \\ 0 \end{array}\right].$$

Let us construct the alias-component matrix $\mathbf{H}(z)$, i.e.,

$$\mathbf{H}(z) = \left[\begin{array}{cc} 1 & z^{-1} \\ 1 & -z^{-1} \end{array}\right].$$

Solving for $\mathbf{H}^{-1}(z)$ yields

$$\mathbf{H}^{-1}(z) = \frac{1}{2} \left[\begin{array}{cc} 1 & 1 \\ z & -z \end{array}\right]$$

so,

$$\mathbf{f}(z) = 2\mathbf{H}^{-1}(z)\mathbf{A}(z).$$

Substituting $\mathbf{H}^{-1}(z)$ and $\mathbf{A}(z)$ into this equation yields

$$\mathbf{f}(z) = 2\left(\frac{1}{2} \left[\begin{array}{cc} 1 & 1 \\ z & -z \end{array}\right]\right) \left[\begin{array}{c} a(z) \\ 0 \end{array}\right].$$

Solving for $\mathbf{f}(z)$ yields

$$\left[\begin{array}{c} F_0(z) \\ F_1(z) \end{array}\right] = \left[\begin{array}{c} a(z) \\ za(z) \end{array}\right].$$

For perfect reconstruction, let $a(z) = z^{-m_0}$, where $m_0 \geq 1$, then

$$\left[\begin{array}{c} F_0(z) \\ F_1(z) \end{array}\right] = \left[\begin{array}{c} z^{-m_0} \\ z^{-(m_0-1)} \end{array}\right].$$

The resulting filter bank is the delay chain filter bank, which was depicted for $m_0 = 1$ in Section 1.5. But if $a(z) = \frac{1}{2}(z^{-2} + z^{-4})$, then

$$
\begin{bmatrix} F_0(z) \\ F_1(z) \end{bmatrix} = \begin{bmatrix} \frac{1}{2}(z^{-2} + z^{-4}) \\ \frac{1}{2}(z^{-1} + z^{-3}) \end{bmatrix}
$$

would be free from aliasing with the output obtained by successively *pasting* the local average of the odd- and even-numbered input data values.

3.2.2.2   Polyphase formulation of filter banks

Now, we consider the polyphase representation formulation of filter banks. Towards this end, we will expand the *desired result* in terms of polyphase. Then, we will determine the conditions to be placed on this result so as to achieve perfect reconstruction or simply alias cancellation.

Before we begin, we need to establish some notation. If $\mathbf{A}$ is a matrix of constants, then $\mathbf{A}^H$ is the transpose-conjugate of $\mathbf{A}$.

Definition 3.2.2.2. *Let $\mathbf{H}(z)$ and $\mathbf{e}(z)$ be a matrix and a vector, respectively, of the complex variable z. The paraconjugate of $\mathbf{H}(z)$, denoted $\widetilde{\mathbf{H}}(z)$, is defined by*

$$
\widetilde{\mathbf{H}}(z) = \mathbf{H}^H(1/z^*)
$$

and the paraconjugate of $\mathbf{e}(z)$, denoted $\widetilde{\mathbf{e}}(z)$, is defined by

$$
\widetilde{\mathbf{e}}(z) = \mathbf{e}^H(1/z^*).
$$

To illustrate this notion of paraconjugation, consider the following examples if $\mathbf{H}(z)$ is given by

$$
\mathbf{H}(z) = \begin{bmatrix} 1 + (2 + 3j)z^{-1} & 2j \\ 3z^{-1} & 4j + z^{-2} \end{bmatrix}
$$

then

$$
\widetilde{\mathbf{H}}(z) = \begin{bmatrix} 1 + (2 - 3j)z & 3z \\ -2j & -4j + z^2 \end{bmatrix}.
$$

If $\mathbf{e}(z)$ is given by

$$
\mathbf{e}(z) = \begin{bmatrix} 1 & z^{-1} \end{bmatrix}^T
$$

then

$$
\widetilde{\mathbf{e}}(z) = \begin{bmatrix} 1 & z \end{bmatrix}.
$$

Now, recall that the desired result of the filter bank formulation is given by

$$
Y(z) = \left[\frac{1}{M} \sum_{k=0}^{M-1} F_k(z) H_k(z) \right] X(z),
$$

or equivalently in matrix notation

$$Y(z) = \frac{1}{M}\mathbf{f}^T(z)\mathbf{h}(z)X(z),$$

where the synthesis bank is given by

$$\mathbf{f}(z) = [F_0(z),\ F_1(z),\ldots,F_{M-1}(z)]^T,$$

and the analysis bank is given by

$$\mathbf{h}(z) = [H_0(z),\ H_1(z),\ldots,H_{M-1}(z)]^T.$$

Writing $H_k(z), k = 0,\ldots M - 1$, in terms of Type-I polyphase yields

$$H_k(z) = \sum_{m=0}^{M-1} z^{-m}E_{k,m}(z^M) \text{ for } k = 0,\ldots M - 1.$$

This system of equations can be rewritten in matrix notation as

$$\mathbf{h}(z) = \mathbf{E}(z^M)\mathbf{e}_M(z),$$

where the delay vector (called a *delay chain*) $\mathbf{e}_M(z)$ is given by

$$\mathbf{e}_M(z) = [1,\ z^{-1},\ldots,z^{-(M-1)}]^T$$

and the polyphase-component matrix for the analysis bank $\mathbf{E}(z)$ is given by

$$\mathbf{E}(z^M) = \begin{bmatrix} E_{0,0}(z^M) & E_{0,1}(z^M) & \cdots & E_{0,M-1}(z^M) \\ E_{1,0}(z^M) & E_{1,1}(z^M) & \cdots & E_{1,M-1}(z^M) \\ \vdots & \vdots & \ddots & \vdots \\ E_{M-1,0}(z^M) & E_{M-1,1}(z^M) & \cdots & E_{M-1,M-1}(z^M) \end{bmatrix}.$$

Then, writing $F_k(z), k = 0,\ldots M - 1$ in terms of Type-II polyphase yields

$$F_k(z) = \sum_{m=0}^{M-1} z^{-(M-1-m)}R_{m,k}(z^M) \text{ for } k = 0,\ldots M - 1.$$

This system of equations can be written in matrix notation as

$$\mathbf{f}^T(z) = z^{-(M-1)}\widetilde{\mathbf{e}}_M(z)\mathbf{R}(z^M),$$

where the paraconjugation of $\mathbf{e}_M(z)$, denoted $\widetilde{\mathbf{e}}_M(z)$, is given by

$$\widetilde{\mathbf{e}}_M(z) = [1,\ z,\ldots,z^{M-1}]$$

and the polyphase-component matrix for the synthesis bank $\mathbf{R}(z)$ is given by

$$\mathbf{R}(z^M) = \begin{bmatrix} R_{0,0}(z^M) & R_{0,1}(z^M) & \cdots & R_{0,M-1}(z^M) \\ R_{1,0}(z^M) & R_{1,1}(z^M) & \cdots & R_{1,M-1}(z^M) \\ \vdots & \vdots & \ddots & \vdots \\ R_{M-1,0}(z^M) & R_{M-1,1}(z^M) & \cdots & R_{M-1,M-1}(z^M) \end{bmatrix}.$$

Substituting these equations for $\mathbf{h}(z)$ and $\mathbf{f}^T(z)$ into the equation for $Y(z)$ yields

$$Y(z) = \frac{1}{M} \underbrace{z^{-(M-1)}\widetilde{\mathbf{e}}_M(z)\mathbf{R}(z^M)}_{\mathbf{f}^T(z)} \underbrace{\mathbf{E}(z^M)\mathbf{e}_M(z)}_{\mathbf{h}(z)} X(z).$$

Let $\mathbf{P}(z) = \mathbf{R}(z)\mathbf{E}(z)$. Then,

$$Y(z) = \frac{1}{M} z^{-(M-1)}\widetilde{\mathbf{e}}_M(z)\mathbf{P}(z^M)\mathbf{e}_M(z)X(z),$$

or equivalently,

$$Y(z) = T(z)X(z)$$

where the distortion function $T(z)$ is given by

$$T(z) = \frac{1}{M} z^{-(M-1)}\widetilde{\mathbf{e}}_M(z)\mathbf{P}(z^M)\mathbf{e}_M(z).$$

These equations suggest the filter bank in Figure 3.7.

3.2.2.3   Relationship between formulations

As seen by the previous subsections, the study of filter banks can be performed using either alias-component or polyphase matrices. Now let us examine the relationship between these two formulations. The AC matrix $\mathbf{H}(z)$ is given by

$$\mathbf{H}(z) = \begin{bmatrix} H_0(z) & H_1(z) & \cdots & H_{M-1}(z) \\ H_0(zW_M) & H_1(zW_M) & \cdots & H_{M-1}(zW_M) \\ \vdots & \vdots & \ddots & \vdots \\ H_0(zW_M^{M-1}) & H_1(zW_M^{M-1}) & \cdots & H_{M-1}(zW_M^{M-1}) \end{bmatrix}$$

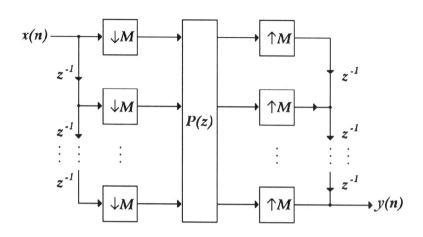

Figure 3.7. Polyphase-component filter bank.

so,

$$\mathbf{H}^T(z) = \begin{bmatrix} H_0(z) & H_0(zW_M) & \cdots & H_0(zW_M^{M-1}) \\ H_1(z) & H_1(zW_M) & \cdots & H_1(zW_M^{M-1}) \\ \vdots & \vdots & \ddots & \vdots \\ H_{M-1}(z) & H_{M-1}(zW_M) & \cdots & H_{M-1}(zW_M^{M-1}) \end{bmatrix}.$$

Recall that $\mathbf{h}(z)$ is given to

$$\mathbf{h}(z) = [H_0(z),\ H_1(z), \ldots, H_{M-1}(z)]^T.$$

Hence,

$$\mathbf{H}^T(z) = [\mathbf{h}(z), \mathbf{h}(zW_M), \ldots, \mathbf{h}(zW_M^{M-1})].$$

Recall that $\mathbf{h}(z)$ is related to the polyphase-component matrix $\mathbf{E}(z)$ by

$$\mathbf{h}(z) = \mathbf{E}(z^M)\mathbf{e}_M(z).$$

Therefore, $\mathbf{H}^T(z)$ can be written as

$$\mathbf{H}^T(z) = [\mathbf{E}(z^M)\mathbf{e}_M(z), \ldots, \mathbf{E}(z^M(W_M^{M-1})^M)\mathbf{e}_M(zW_M^{M-1})],$$

or simply

$$\mathbf{H}^T(z) = \mathbf{E}(z^M)[\mathbf{e}_M(z),\ \mathbf{e}_M(zW_M), \ldots, \mathbf{e}_M(zW_M^{M-1})].$$

But by definition

$$\mathbf{e}_M(zW_M^k) = [1,\; z^{-1}W_M^{-k}, \ldots, z^{-(M-1)}W_M^{-(M-1)k}]^T,$$

or equivalently,

$$\mathbf{e}_M(zW_M^k) = \mathbf{\Lambda}(z) \begin{bmatrix} 1 \\ W_M^{-k} \\ \vdots \\ W_M^{-(M-1)k} \end{bmatrix}$$

where $\mathbf{\Lambda}(z) = \mathrm{diag}[1,\; z^{-1}, \ldots, z^{-(M-1)}]$. Note that

$$\begin{bmatrix} 1 \\ W_M^{-k} \\ \vdots \\ W_M^{-(M-1)k} \end{bmatrix}$$

is just a single column of the $\mathbf{W}_M^H$ matrix, where $\mathbf{W}_M$ is the M x M discrete Fourier transform matrix. So,

$$\mathbf{H}^T(z) = \mathbf{E}(z^M)\mathbf{\Lambda}(z)\mathbf{W}_M^H.$$

Since $\mathbf{W}_M^H = (\mathbf{W}_M^H)^T$, then

$$\mathbf{H}(z) = \mathbf{W}_M^H \mathbf{\Lambda}(z)\mathbf{E}^T(z^M).$$

With this equation, any results obtained in terms of the polyphase formulation can be applied to the alias-component formulation and vice versa.

3.2.3   A multirate source-coding design example

We will now apply the results of the last subsection to a multirate source-coding design problem. Let $x(n)$ be an arbitrary sequence and let $x_1(n)$ be its first divided difference, *i.e.*

$$x_1(n) = x(n) - x(n-1).$$

Consider two sequences $y_0(n)$ and $y_1(n)$ that are defined by

$$y_0(n) = x(2n) \text{ and } y_1(n) = x_1(2n).$$

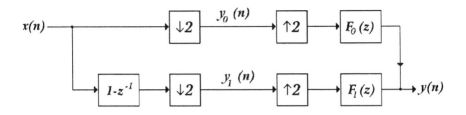

Figure 3.8. Filter bank equivalent circuit.

Can we recover $x(n)$ from $y_0(n)$ and $y_1(n)$? It is important to note that the even-numbered samples are already available. Can the odd-numbered samples be recovered? To solve this problem, let us first recast it as a two-channel QMF problem, where the analysis filters are $H_0(z) = 1$ and $H_1(z) = 1 - z^{-1}$, the z-transform domain expression for the first divided difference. So the filter bank is depicted in Figure 3.8, where the goal is to determine $F_0(z)$ and $F_1(z)$ such that we have perfect reconstruction. Let us first determine the polyphase matrix $\mathbf{E}(z)$. From the last subsection, we found that

$$\mathbf{H}(z) = \mathbf{W}_2^H \mathbf{\Lambda}(z) \mathbf{E}^T(z^2).$$

Recall that $\mathbf{W}_2 \mathbf{W}_2^H = 2\mathbf{I}_2$ and $\mathbf{\Lambda}(z)\widetilde{\mathbf{\Lambda}}(z) = \mathbf{I}_2$. So, solving for $\mathbf{E}^T(z^2)$ we obtain

$$\mathbf{E}^T(z^2) = \widetilde{\mathbf{\Lambda}}(z)(\frac{1}{2}\mathbf{W}_2)\mathbf{H}(z)$$

and substituting the appropriate entries into the matrices results in

$$\mathbf{E}^T(z^2) = \left[\begin{array}{cc} 1 & 0 \\ 0 & z \end{array} \right] \left(\frac{1}{2} \left[\begin{array}{cc} 1 & 1 \\ 1 & -1 \end{array} \right] \right) \left[\begin{array}{cc} 1 & 1 - z^{-1} \\ 1 & 1 + z^{-1} \end{array} \right],$$

or simply

$$\mathbf{E}^T(z^2) = \left[\begin{array}{cc} 1 & 1 \\ 0 & -1 \end{array} \right].$$

Hence,

$$\mathbf{E}(z) = \left[\begin{array}{cc} 1 & 0 \\ 1 & -1 \end{array} \right].$$

Now, choose $\mathbf{R}(z) = \mathbf{E}^{-1}(z)$ to insure perfect reconstruction, $i.e.$ $y(n) = x(n-1)$. Then,

$$\mathbf{R}(z) = \left[\begin{array}{cc} 1 & 0 \\ 1 & -1 \end{array} \right].$$

Since the entries of $\mathbf{R}(z)$ are not functions of z, then $\mathbf{R}(z^2) = \mathbf{R}(z)$. Therefore,

$$\mathbf{R}(z^2) = \begin{bmatrix} 1 & 0 \\ 1 & -1 \end{bmatrix}.$$

Compute the synthesis filters using

$$\mathbf{f}^T(z) = z^{-1}\widetilde{\mathbf{e}}_M(z)\mathbf{R}(z^2),$$

and substituting entries into the matrices results in

$$\begin{bmatrix} F_0(z) & F_1(z) \end{bmatrix} = z^{-1} \begin{bmatrix} 1 & z \end{bmatrix} \begin{bmatrix} 1 & 0 \\ 1 & -1 \end{bmatrix},$$

or equivalently,

$$\begin{bmatrix} F_0(z) & F_1(z) \end{bmatrix} = \begin{bmatrix} 1 + z^{-1} & -1 \end{bmatrix}.$$

Hence,

$$F_0(z) = 1 + z^{-1} \text{ and } F_1(z) = -1.$$

Assume $x(n)$ is a slowly varying sequence, that is, adjacent samples differ by a small amount. If each sample of $x(n)$ requires 16 bits for its representation, then the first divided difference, being very small, requires only 8 bits for its representation. Now instead of storing (or transmitting) all samples of $x(n)$, we store (or transmit) $y_0(n)$ using 16 bits per sample and $y_1(n)$ using 8 bits per sample. Thus, we have reduced the data rate from 16 bits per sample to an average of 12 bits per sample.

Note, however, that if the polyphase matrices are utilized directly, that is,

$$\mathbf{E}(z) = \mathbf{R}(z) = \begin{bmatrix} 1 & 0 \\ 1 & -1 \end{bmatrix},$$

then the filter bank can be operated at half the rate. The resulting filter bank is depicted in Figure 3.9. The regular structure of this realization foreshadows the discussion of lattice structures in the next chapter.

3.2.4   Quantization effects and filter banks

If the filter bank is used for compression, then there will be information loss between the analysis bank and the synthesis bank. This loss is due to coding errors that are the result of quantization effects. This subsection will show that if the quantizer noise is correlated with the input signal as in a Lloyd-Max quantizer, then a slight modification to a filter bank can result in the cancellation of all signal-dependent errors.

For completeness, we will briefly review Leibnitz's rule and continuous random variables, since they will be used in the derivation of the Lloyd-Max quantizer.

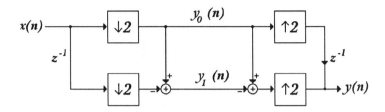

Figure 3.9. Efficient realization of design problem.

3.2.4.1 Leibnitz's rule

Leibnitz's rule describes how we can differentiate a function which is of the form $\int_{\nu_1(t)}^{\nu_2(t)} g(x,t)dx$.

Theorem 3.2.4.1. *Let*

$$\beta(t) = \int_{\nu_1(t)}^{\nu_2(t)} g(x,t)dx$$

where ν_1 and ν_2 may depend on the parameter t and $a \le t \le b$. Then,

$$\frac{d\beta}{dt} = \int_{\nu_1(t)}^{\nu_2(t)} \frac{\partial g}{\partial t}dx + g(\nu_2,t)\frac{d\nu_2}{dt} - g(\nu_1,t)\frac{d\nu_1}{dt}$$

for $a \le t \le b$, if $g(x,t)$ and $\frac{\partial g}{\partial t}$ are continuous in both x and t in some region of the xt plane including $\nu_1 \le x \le \nu_2$, $a \le t \le b$ and if ν_1 and ν_2 are continuous and have continuous derivatives for $a \le t \le b$.

Proof: Let

$$\beta(t) = \int_{\nu_1(t)}^{\nu_2(t)} g(x,t)dx.$$

Let $t \in (a,b)$ and $\Delta t \ne 0$ such that $t + \Delta t \in (a,b)$, then

$$\Delta\beta = \beta(t+\Delta t) - \beta(t)$$

becomes

$$\Delta\beta = \int_{\nu_1(t+\Delta t)}^{\nu_2(t+\Delta t)} g(x,t+\Delta t)dx - \int_{\nu_1(t)}^{\nu_2(t)} g(x,t)dx,$$

or equivalently,

$$\Delta\beta = -\int_{\nu_1(t+\Delta t)}^{\nu_1(t)} g(x,t+\Delta t)dx + \int_{\nu_1(t)}^{\nu_2(t)} g(x,t+\Delta t)dx$$
$$+ \int_{\nu_2(t)}^{\nu_2(t+\Delta t)} g(x,t+\Delta t)dx - \int_{\nu_1(t)}^{\nu_2(t)} g(x,t)dx.$$

Now, grouping the second and fourth terms yields

$$\Delta\beta = \int_{\nu_1(t)}^{\nu_2(t)} \left[g(x,t+\Delta t) - g(x,t)\right] dx$$
$$+ \int_{\nu_2(t)}^{\nu_2(t+\Delta t)} g(x,t+\Delta t)dx - \int_{\nu_1(t)}^{\nu_1(t+\Delta t)} g(x,t+\Delta t)dx.$$

Now scale this equation by Δt yields

$$\frac{\Delta\beta}{\Delta t} = \int_{\nu_1(t)}^{\nu_2(t)} \frac{[g(x,t+\Delta t)-g(x,t)]}{\Delta t} dx$$
$$+ \frac{1}{\Delta t}\int_{\nu_2(t)}^{\nu_2(t+\Delta t)} g(x,t+\Delta t)dx - \frac{1}{\Delta t}\int_{\nu_1(t)}^{\nu_1(t+\Delta t)} g(x,t+\Delta t)dx.$$

Applying the mean-value theorem for integrals to the last two terms yields

$$\frac{\Delta\beta}{\Delta t} = \int_{\nu_1(t)}^{\nu_2(t)} \frac{[g(x,t+\Delta t)-g(x,t)]}{\Delta t} dx$$
$$+ g(\xi_2,t+\Delta t)\frac{[\nu_2(t+\Delta t)-\nu_2(t)]}{\Delta t}$$
$$- g(\xi_1,t+\Delta t)\frac{[\nu_1(t+\Delta t)-\nu_1(t)]}{\Delta t},$$

where ξ_2 is between $\nu_2(t)$ and $\nu_2(t+\Delta t)$ and ξ_1 is between $\nu_1(t)$ and $\nu_1(t+\Delta t)$. Then, taking the limit as $\Delta t \to 0$, the equation becomes

$$\frac{d\beta}{dt} = \int_{\nu_1}^{\nu_2} \frac{\partial g}{\partial t}dx + g(\nu_2,t)\frac{d\nu_2}{dt} - g(\nu_1,t)\frac{d\nu_1}{dt}.$$

∎

3.2.4.2 Continuous random variables

Suppose X is a continuous random variable on the interval with a probability density function f. Then, the probability that $a \leq X \leq b$ is given by

$$P(a \leq X \leq b) = \int_a^b f(x)dx$$

where, f is called the probability density function. The *expected value* of X on the interval $[a,b]$, denoted $E[X]$, is given by

$$E[X] = \frac{\int_a^b xf(x)dx}{\int_a^b f(x)dx}$$

where $a \leq X \leq b$. Moreover, for any positive integer k,

$$E[X^k] = \frac{\int_a^b x^k f(x)dx}{\int_a^b f(x)dx}$$

where $a \leq X \leq b$. The *variance* of X, denoted σ_X^2, is given by

$$\sigma_X^2 = E\left[X - E[X]\right]^2.$$

Substituting the definition of the expected value yields

$$\sigma_X^2 = \frac{\int_a^b \left[x - E[X]\right]^2 f(x)dx}{\int_a^b f(x)dx},$$

or equivalently,

$$\sigma_X^2 = \frac{\int_a^b x^2 f(x)dx - 2E[X]\int_a^b xf(x)dx + [E[X]]^2 \int_a^b f(x)dx}{\int_a^b f(x)dx}.$$

Simplifying this equation using the definition of $E[X]$ and $E[X^2]$ yields

$$\sigma_X^2 = \frac{E[X^2]\int_a^b f(x)dx - 2[E[X]]^2 \int_a^b f(x)dx + [E[X]]^2 \int_a^b f(x)dx}{\int_a^b f(x)dx},$$

or simply,

$$\sigma_X^2 = E[X^2] - [E[X]]^2.$$

3.2.4.3   Lloyd-Max quantizers

For completeness, the following description of Lloyd-Max quantizers [Max, 1960] is included. These quantizers minimize the expected value of the distortion (or error between the input and the output of the quantizer) of a signal by a quantizer when the number of levels of the quantizer is fixed.

Let u be a random variable with a continuous probability density function $p(u)$. It is desired to find the set of increasing transition or decision levels $\{t_m \mid m = 1, \ldots L\}$ with t_1 and t_L as the minimum and maximum values of u, respectively. If u lies in the interval $[t_m, t_{m+1})$, then it is mapped to r_m, the mth reconstruction level. The Lloyd-Max quantizer minimizes the mean square error for a given number of quantization levels. Therefore, for an L-level quantizer, the mean square error

$$\mathcal{E} = \sum_{k=1}^{L-1} \int_{t_k}^{t_{k+1}} (u - r_k)^2 p(u)du$$

is to be minimized. The necessary conditions for this minimization are obtained by differentiating $\mathcal{E}$ with respect to t_m and r_m and setting the result equal to zero.

Now let us consider $\frac{\partial \mathcal{E}}{\partial r_m}$. It yields

$$\frac{\partial \mathcal{E}}{\partial r_m} = \frac{\partial}{\partial r_m} \sum_{k=1}^{L-1} \int_{t_k}^{t_{k+1}} (u - r_k)^2 p(u) du.$$

By inspection, this partial derivative is only nonzero for $k = m$. Hence,

$$\frac{\partial \mathcal{E}}{\partial r_m} = \frac{\partial}{\partial r_m} \int_{t_m}^{t_{m+1}} (u - r_m)^2 p(u) du,$$

or equivalently,

$$\frac{\partial \mathcal{E}}{\partial r_m} = -2 \int_{t_m}^{t_{m+1}} (u - r_m) p(u) du.$$

Setting $\frac{\partial \mathcal{E}}{\partial r_m} = 0$ yields

$$\int_{t_m}^{t_{m+1}} (u - r_m) p(u) du = 0,$$

or equivalently,

$$\int_{t_m}^{t_{m+1}} u p(u) du = r_m \int_{t_m}^{t_{m+1}} p(u) du.$$

So,

$$r_m = \frac{\int_{t_m}^{t_{m+1}} u p(u) du}{\int_{t_m}^{t_{m+1}} p(u) du}.$$

Thus, the reconstruction levels of the Lloyd-Max quantizer lie at the center of mass of the probability density between transition levels. Using the definition of expected value of a continuous random variable, we can reinterpret r_m as

$$r_m = E[U]$$

where U is a random variable and $t_m \leq U \leq t_{m+1}$.

Now let us consider $\frac{\partial \mathcal{E}}{\partial t_m}$. It yields

$$\frac{\partial \mathcal{E}}{\partial t_m} = \frac{\partial}{\partial t_m} \sum_{k=1}^{L-1} \int_{t_k}^{t_{k+1}} (u - r_k)^2 p(u) du.$$

By inspection, this partial derivative is only nonzero for $k = m - 1$ and $k = m$. Hence,

$$\frac{\partial \mathcal{E}}{\partial t_m} = \frac{\partial}{\partial t_m} \left[\int_{t_{m-1}}^{t_m} (u - r_{m-1})^2 p(u) du + \int_{t_m}^{t_{m+1}} (u - r_m)^2 p(u) du \right].$$

Then, using Leibnitz's rule, we obtain

$$\frac{\partial \mathcal{E}}{\partial t_m} = (t_m - r_{m-1})^2 p(t_m) - (t_m - r_m)^2 p(t_m).$$

Setting $\frac{\partial \mathcal{E}}{\partial t_m} = 0$ requires us to set to zero the coefficient of $p(t_m)$. Thus,

$$(t_m - r_{m-1})^2 = (t_m - r_m)^2,$$

or equivalently,

$$t_m^2 - 2t_m r_{m-1} + r_{m-1}^2 = t_m^2 - 2t_m r_m + r_m^2.$$

Hence,

$$2t_m(r_m - r_{m-1}) = r_m^2 - r_{m-1}^2,$$

or equivalently,

$$t_m = \frac{r_m + r_{m-1}}{2}.$$

Thus, for the Lloyd-Max quantizer, the transition levels lie halfway between the reconstruction levels. We can reinterpret t_m as

$$t_m = E[X]$$

where X is a uniformly distributed random variable and $r_{m-1} \leq X \leq r_m$.

3.2.4.4 Quantizer models

As we just observed, the Lloyd-Max quantizer is an example of a quantizer, whose quantization noise is correlated with the input signal. The most obvious model for such a quantizer is given by Figure 3.10, or equivalently,

$$s(n) = t(n) + q(n),$$

where the input signal $t(n)$ and the additive noise $q(n)$ are correlated. We will show that we could equivalently use the *gain plus additive noise* model, which can be expressed as

$$s(n) = \alpha t(n) + r(n)$$

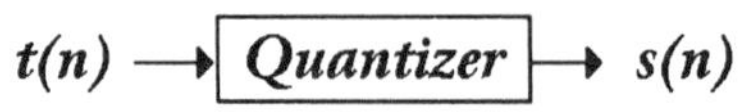

$$t(n) \longrightarrow \boxed{Quantizer} \longrightarrow s(n)$$

Figure 3.10. First quantizer model.

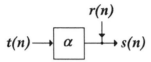

Figure 3.11. Second quantizer model.

or equivalently in block diagram form in Figure 3.11, where α is the gain factor ($0 < \alpha \leq 1$) and $r(n)$ is the additive noise term. By making the proper choice of the gain α, we will be able to assume that $r(n)$ is additive uncorrelated noise. For these quantizers, assume that the expected value of the quantization noise $q(n)$ is zero, that is,

$$E[q(n)] = 0.$$

Let us briefly examine some interrelationships between the two models. From the first model,

$$E[s(n)] = E[t(n) + q(n)],$$

or equivalently,

$$E[s(n)] = E[t(n)] + E[q(n)].$$

Since $E[q(n)] = 0$, then

$$E[s(n)] = E[t(n)].$$

From the second model,

$$E[s(n)] = E[\alpha t(n) + r(n)],$$

or equivalently,

$$E[s(n)] = \alpha E[t(n)] + E[r(n)].$$

Equating the two equations for $E[s(n)]$ yields

$$(1 - \alpha)E[t(n)] = E[r(n)].$$

Now, assume that the quantization noise $q(n)$ and the quantizer output $s(n)$ are independent uncorrelated processes. Then,

$$E[q(n)s(n)] = E[q(n)]E[s(n)] = 0.$$

Using this property, we will examine $E[t(n)q(n)]$ for the first model.

$$
\begin{aligned}
E[t(n)q(n)] &= E[(s(n) - q(n))q(n)] \\
&= E[s(n)q(n)] - E[q^2(n)] \\
&= -\sigma_q^2.
\end{aligned}
$$

Using the second model, we obtain the following.

$$
\begin{aligned}
E[t(n)q(n)] &= E[t(n)(s(n) - t(n))] \\
&= E[t(n)(\alpha t(n) + r(n) - t(n))] \\
&= (\alpha - 1)E[t^2(n)] + E[t(n)r(n)] \\
&= (\alpha - 1)\sigma_t^2 + (\alpha - 1)(E[t(n)])^2 + E[t(n)r(n)].
\end{aligned}
$$

Since $(1 - \alpha)E[t(n)] = E[r(n)]$, then $E[t(n)q(n)]$ becomes

$$E[t(n)q(n)] = (\alpha - 1)\sigma_t^2 - E[t(n)]E[r(n)] + E[t(n)r(n)].$$

Equating the two equations for $E[t(n)q(n)]$ yields

$$-\sigma_q^2 = (\alpha - 1)\sigma_t^2 - E[t(n)]E[r(n)] + E[t(n)r(n)],$$

or equivalently,

$$E[t(n)r(n)] - E[t(n)]E[r(n)] = (1 - \alpha)\sigma_t^2 - \sigma_q^2.$$

Now, if we choose α to be

$$\alpha = 1 - \frac{\sigma_q^2}{\sigma_t^2},$$

then the input signal $t(n)$ and the additive noise $r(n)$ are uncorrelated processes, that is,

$$E[t(n)r(n)] - E[t(n)]E[r(n)] = 0.$$

Let us examine $E[s^2(n)]$ for the two models. From the first model,

$$E[s^2(n)] = E\left[(t(n) + q(n))^2\right],$$

or equivalently,

$$E[s^2(n)] = E[t^2(n)] + 2E[t(n)q(n)] + E[q^2(n)].$$

Since $E[t(n)q(n)] = (\alpha - 1)\sigma_t^2$ and $E[q^2(n)] = \sigma_q^2 = (1 - \alpha)\sigma_t^2$, then

$$E[s^2(n)] = E[t^2(n)] + 2(\alpha - 1)\sigma_t^2 + (1 - \alpha)\sigma_t^2,$$

or equivalently,

$$E[s^2(n)] = E[t^2(n)] + (\alpha - 1)\sigma_t^2.$$

Since $E[t^2(n)] = \sigma_t^2 + (E[t(n)])^2$, then

$$E[s^2(n)] = \alpha\sigma_t^2 + (E[t(n)])^2.$$

From the second model,

$$E[s^2(n)] = E\left[(\alpha t(n) + r(n))^2\right],$$

or equivalently,

$$E[s^2(n)] = \alpha^2 E[t^2(n)] + 2\alpha E[t(n)r(n)] + E[r^2(n)].$$

Since $E[t(n)r(n)] = E[t(n)]E[r(n)]$, then

$$E[s^2(n)] = \alpha^2 E[t^2(n)] + 2\alpha E[t(n)]E[r(n)] + E[r^2(n)].$$

Since $E[t^2(n)] = \sigma_t^2 + (E[t(n)])^2$ and $E[r^2(n)] = \sigma_r^2 + (E[r(n)])^2$, then

$$E[s^2(n)] = \alpha^2 \left(\sigma_t^2 + (E[t(n)])^2\right) + 2\alpha E[t(n)]E[r(n)] + \sigma_r^2 + (E[r(n)])^2.$$

Since $E[r(n)] = (1 - \alpha)E[t(n)]$, then

$$E[s^2(n)] = \alpha^2\sigma_t^2 + \sigma_r^2 + (\alpha^2 + 2\alpha(1 - \alpha) + (1 - \alpha)^2)(E[t(n)])^2$$

or equivalently,

$$E[s^2(n)] = \alpha^2\sigma_t^2 + \sigma_r^2 + (E[t(n)])^2.$$

Equating the two equations for $E[s^2(n)]$ yields

$$\alpha\sigma_t^2 + (E[t(n)])^2 = \alpha^2\sigma_t^2 + \sigma_r^2 + (E[t(n)])^2,$$

or equivalently,

$$\sigma_r^2 = \alpha(1 - \alpha)\sigma_t^2.$$

Thus, the *gain plus additive noise* model is characterized by gain $\alpha = 1 - \frac{\sigma_q^2}{\sigma_t^2}$ and a noise source $r(n)$, which is characterized by $E[r(n)] = (1 - \alpha)E[t(n)]$ and $\sigma_r^2 = \alpha(1 - \alpha)\sigma_t^2$.

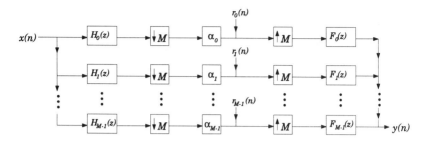

Figure 3.12. M-channel filter bank with quantizers.

Figure 3.13. Additive noise equivalent circuit.

3.2.4.5 Filter bank with quantizers

The filter bank with quantizers will consist of the insertion of the quantizer model in each channel between the analysis bank and the synthesis bank. This is depicted by the block diagram in Figure 3.12.

The basic philosophy behind the design of QMF banks with quantizers is to permit aliasing in the filters of the analysis bank and coding gain in the quantizers, and then to choose the filters of the synthesis bank so that the alias-components in the filters of the analysis bank and the coding gain of the quantizers are cancelled.

The analysis of the QMF bank can be performed easily. First, realize that the effect of α_k is simply a scale factor for both the desired terms and the terms due to aliasing. Second, the effect of the additive noise source $r_k(n)$ can be described by the equivalent circuit per channel as depicted in Figure 3.13. Thus, the filter bank with quantizers yields

$$
\begin{aligned}
Y(z) \quad = \quad & [\tfrac{1}{M} \textstyle\sum_{k=0}^{M-1} \alpha_k F_k(z) H_k(z)] X(z) \\
& + \textstyle\sum_{n=1}^{M-1} [\tfrac{1}{M} \textstyle\sum_{k=0}^{M-1} \alpha_k F_k(z) H_k(z W_M^n)] X(z W_M^n) \\
& + \textstyle\sum_{k=0}^{M-1} F_k(z) R_k(z^M).
\end{aligned}
$$

Now choose new synthesis filters $G_k(z)$ to be defined by

$$
G_k(z) = \alpha_k F_k(z).
$$

Then, the equation for $Y(z)$ becomes

$$Y(z) \quad = \quad \underbrace{[\frac{1}{M} \sum_{k=0}^{M-1} G_k(z) H_k(z)] X(z)}_{\text{desired terms}}$$

$$+ \quad \underbrace{\sum_{n=1}^{M-1} [\frac{1}{M} \sum_{k=0}^{M-1} G_k(z) H_k(z W_M^n)] X(z W_M^n)}_{\text{terms due to aliasing}}$$

$$+ \quad \underbrace{\sum_{k=0}^{M-1} \frac{G_k(z) R_k(z^M)}{\alpha_k}}_{\text{terms due to quantization}} .$$

Therefore, the errors due to quantization are random signal-independent errors, which could be eliminated by post processing the output of the filter bank with a filter that is designed to remove additive uncorrelated noise. In addition, the analysis that was presented in the previous subsections is still applicable for all signal-dependent terms.

3.3   Foundations of filter banks

In the previous section, we showed that the polyphase matrix $\mathbf{P}(z)$ charac-terizes the behavior of a QMF bank. What are the properties of $\mathbf{P}(z)$ that correspond to alias-free or perfect reconstruction filter banks? This section will address these questions.

3.3.1   Alias-free filter banks

First, a background section on pseudocirculant matrices is presented. This is followed by the theory of alias-free filter banks.

3.3.1.1   Pseudocirculant matrices

Definition 3.3.1.1. *A matrix* $\mathbf{P}(z)$ *is said to be pseudocirculant if it is of the following form*

$$\mathbf{P}(z) = \begin{bmatrix} P_0(z) & P_1(z) & \cdots & P_{M-1}(z) \\ z^{-1}P_{M-1}(z) & P_0(z) & \cdots & P_{M-2}(z) \\ \vdots & \vdots & \ddots & \vdots \\ z^{-1}P_1(z) & z^{-1}P_2(z) & \cdots & P_0(z) \end{bmatrix}.$$

Two important special cases of the pseudocirculant property are given below.

(1) If $z = 1$, then $\mathbf{P}(z)$ is circulant.

(2) If $z = -1$, then $\mathbf{P}(z)$ is skew circulant.

For example, a pseudocirculant matrix $\mathbf{P}(z)$ of dimension three is given by

$$\mathbf{P}(z) = \begin{bmatrix} P_0(z) & P_1(z) & P_2(z) \\ z^{-1}P_2(z) & P_0(z) & P_1(z) \\ z^{-1}P_1(z) & z^{-1}P_2(z) & P_0(z) \end{bmatrix}.$$

3.3.1.2 Eliminating alias distortion

Consider the filter bank depicted in Figure 3.14.

Now, we will proceed with the analysis of the filter bank, by examining it stage-by-stage. First, we will consider, the bank of decimators, which is depicted in Figure 3.15. The elemental equation for this stage is given by

$$C_n(z) = \frac{1}{M} \sum_{k=0}^{M-1} (z^{1/M} W_M^k)^{-n} X(z^{1/M} W_M^k).$$

Next, we will consider the $\mathbf{P}(z)$ block, which is depicted in Figure 3.16. The elemental equation for this stage is given by

$$B_s(z) = \sum_{n=0}^{M-1} P_{s,n}(z) C_n(z).$$

Lastly, we consider the expander block as given in Figure 3.17. The elemental equation for this stage is given by

$$Y(z) = \sum_{s=0}^{M-1} z^{-(M-1-s)} B_s(z^M).$$

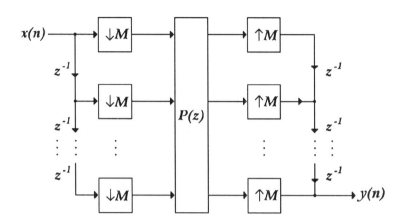

Figure 3.14. Polyphase representation of M-channel filter bank.

Combining the equation for $C_n(z)$ and $B_s(z)$ yields

$$B_s(z) = \sum_{n=0}^{M-1} P_{s,n}(z) \left[\frac{1}{M} \sum_{k=0}^{M-1} (z^{1/M} W_M^k)^{-n} X(z^{1/M} W_M^k) \right].$$

Combining this equation with the one for $Y(z)$ yields

$$Y(z) = \sum_{s=0}^{M-1} z^{-(M-1-s)} \sum_{n=0}^{M-1} P_{s,n}(z^M) \left[\frac{1}{M} \sum_{k=0}^{M-1} (z W_M^k)^{-n} X(z W_M^k) \right],$$

or equivalently,

$$Y(z) = \frac{1}{M} \sum_{k=0}^{M-1} X(z W_M^k) \sum_{n=0}^{M-1} W_M^{-kn} V_n(z)$$

where

$$V_n(z) = \sum_{s=0}^{M-1} z^{-n} z^{-(M-1-s)} P_{s,n}(z^M).$$

Since $X(z W_M^k)$, $k \neq 0$, represents the alias terms, then the resulting equation for $Y(z)$ is free from aliasing if and only if the coefficients of $X(z W_M^k)$, $k \neq 0$, are zero, that is,

$$\sum_{n=0}^{M-1} W_M^{-kn} V_n(z) = 0 \text{ for } k \neq 0,$$

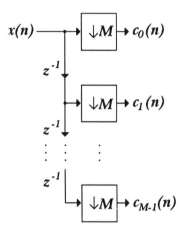

Figure 3.15. Bank of decimator.

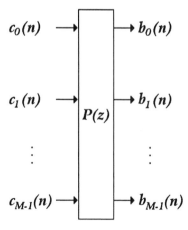

Figure 3.16. $\mathbf{P}(z)$ block.

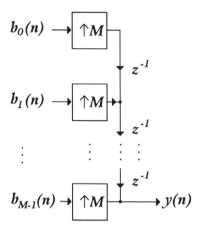

Figure 3.17. Bank of expanders.

or equivalently in matrix notation,

$$\mathbf{W}_M^H \begin{bmatrix} V_0(z) \\ V_1(z) \\ \vdots \\ V_{M-1}(z) \end{bmatrix} = \begin{bmatrix} * \\ 0 \\ \vdots \\ 0 \end{bmatrix}$$

where $*$ is a nonzero term. Premultiply both sides of this equation by $\mathbf{W}_M$. Then, using the fact that $\mathbf{W}_M \mathbf{W}_M^H = M\mathbf{I}$, this equation becomes

$$M \begin{bmatrix} V_0(z) \\ V_1(z) \\ \vdots \\ V_{M-1}(z) \end{bmatrix} = \mathbf{W}_M \begin{bmatrix} * \\ 0 \\ \vdots \\ 0 \end{bmatrix}.$$

Since all the entries in the first column of $\mathbf{W}_M$ are equal to one, then

$$V_0(z) = V_1(z) = \cdots = V_{M-1}(z).$$

Therefore, let

$$V(z) = V_n(z) \text{ for } n = 0, \ldots, M - 1$$

and the alias terms become

$$\frac{1}{M} \sum_{k=1}^{M-1} X(zW_M^k) \sum_{n=0}^{M-1} W_M^{-kn} V_n(z) = \frac{1}{M} \sum_{k=1}^{M-1} X(zW_M^k) V(z) \sum_{n=0}^{M-1} W_M^{-kn}$$

but

$$\sum_{n=0}^{M-1} W_M^{-kn} = \sum_{n=0}^{M-1} \exp(\frac{j2\pi kn}{M}) = \frac{1 - \exp(j2\pi k)}{1 - \exp(\frac{j2\pi k}{M})} = 0 \text{ for } k \neq 0.$$

Thus,

$$Y(z) = V(z)X(z)$$

and the M-channel filter bank is free from aliasing. Now, let us examine $V_0(z)$ and $V_1(z)$ more carefully. By definition, $V_0(z)$ is given by

$$V_0(z) = \sum_{s=0}^{M-1} z^{-(M-1-s)} P_{s,0}(z^M).$$

Perform a change of variables by letting $q = M - 1 - s$. Then,

$$V_0(z) = \sum_{q=0}^{M-1} z^{-q} P_{M-1-q,0}(z^M).$$

Similarly, by definition $V_1(z)$ is given by

$$V_1(z) = \sum_{s=0}^{M-1} z^{-1} z^{-(M-1-s)} P_{s,1}(z^M),$$

or equivalently,

$$V_1(z) = z^{-M} P_{0,1}(z^M) + \sum_{s=1}^{M-1} z^{-(M-s)} P_{s,1}(z^M).$$

Perform a change of variables by letting $q = M - s$. Then,

$$V_1(z) = z^{-M} P_{0,1}(z^M) + \sum_{q=1}^{M-1} z^{-q} P_{M-q,1}(z^M).$$

Because of the requirement that $V_0(z) = V_1(z)$, the polyphase-components should be the same. So, $V_0(z)$ is simply a pseudocirculant shift of $V_1(z)$. Similarly, the nth column is obtained from the $(n+1)$st. Thus, $\mathbf{P}(z)$ must be pseudocirculant for the filter bank to be alias-free.

Example 3.3.1.1. Consider the following example that illustrates these results. Consider the following $\mathbf{P}(z)$ matrix

$$\mathbf{P}(z) = \begin{bmatrix} P_{00}(z) & P_{01}(z) & P_{02}(z) \\ P_{10}(z) & P_{11}(z) & P_{12}(z) \\ P_{20}(z) & P_{21}(z) & P_{22}(z) \end{bmatrix}.$$

Now,

$$V_n(z) = \sum_{s=0}^{M-1} z^{-n} z^{-(M-1-s)} P_{s,n}(z^M).$$

Thus,

$$\begin{aligned} V_0(z) &= z^{-2} P_{00}(z^3) &+& z^{-1} P_{10}(z^3) &+& P_{20}(z^3) \\ V_1(z) &= z^{-3} P_{01}(z^3) &+& z^{-2} P_{11}(z^3) &+& z^{-1} P_{21}(z^3) \\ V_2(z) &= z^{-4} P_{02}(z^3) &+& z^{-3} P_{12}(z^3) &+& z^{-2} P_{22}(z^3). \end{aligned}$$

Since

$$V_0(z) = V_1(z) = V_2(z) = V(z)$$

then

$$\begin{aligned} P_{11}(z) &= P_{22}(z) \\ P_{01}(z) &= P_{12}(z) \\ P_{00}(z) &= P_{11}(z) \\ P_{10}(z) &= P_{21}(z) \\ P_{20}(z) &= z^{-1} P_{01}(z) \\ P_{21}(z) &= z^{-1} P_{02}(z). \end{aligned}$$

Hence, we can write

$$\mathbf{P}(z) = \begin{bmatrix} P_{22}(z) & P_{12}(z) & P_{02}(z) \\ z^{-1} P_{02}(z) & P_{22}(z) & P_{12}(z) \\ z^{-1} P_{12}(z) & z^{-1} P_{02}(z) & P_{22}(z) \end{bmatrix}.$$

Let $P_2(z) = P_{02}(z)$, $P_1(z) = P_{12}(z)$, and $P_0(z) = P_{22}(z)$. Then, $\mathbf{P}(z)$ becomes

$$\mathbf{P}(z) = \begin{bmatrix} P_0(z) & P_1(z) & P_2(z) \\ z^{-1} P_2(z) & P_0(z) & P_1(z) \\ z^{-1} P_1(z) & z^{-1} P_2(z) & P_0(z) \end{bmatrix}.$$

3.3.2 Perfect reconstruction filter banks

First, a background section on paraunitary matrices is presented. This is followed by the theory of perfect reconstruction filter banks.

3.3.2.1 Paraunitary matrices

Definition 3.3.2.1. *A matrix* $\mathbf{P}(z)$ *is said to be paraunitary if and only if*

$$\widetilde{\mathbf{P}}(z)\mathbf{P}(z) = c\mathbf{I}, \text{ for all } z,$$

where c is a positive constant. If $c = 1$, *then* $\mathbf{P}(z)$ *known as a normalized paraunitary matrix.*

Two important special cases of the paraunitary property are given below when $\mathbf{P}(z)$ is a constant matrix.

(1) If $\mathbf{W}$ is a matrix of complex-valued constants,

then $\mathbf{W}$ is *unitary* if and only if $\mathbf{W}^H\mathbf{W} = \mathbf{I}$.

(2) If $\mathbf{U}$ is a matrix of real-valued constants,

then $\mathbf{U}$ is *orthogonal* if and only if $\mathbf{U}^T\mathbf{U} = \mathbf{I}$.

Let us consider the following theorem, which provides a fundamental result for products of paraconjugated matrices.

Theorem 3.3.2.1. *Let* $\mathbf{A}(z), \mathbf{B}(z)$, *and* $\mathbf{C}(z)$ *be matrices such that*

$$\mathbf{A}(z) = \mathbf{C}(z)\mathbf{B}(z),$$

then $\widetilde{\mathbf{A}}(z)=\widetilde{\mathbf{B}}(z)\widetilde{\mathbf{C}}(z)$.

Proof: Let $\mathbf{A}(z)$ be l x n, $\mathbf{C}(z)$ be l x m and $\mathbf{B}(z)$ be m x n so that $\mathbf{C}(z)\mathbf{B}(z)$ is defined. Then, $\widetilde{\mathbf{C}}(z)$ is m x l and $\widetilde{\mathbf{B}}(z)$ is n x m so $\widetilde{\mathbf{B}}(z)\widetilde{\mathbf{C}}(z)$ is also defined and it has the same dimension as $\widetilde{\mathbf{A}}(z)$.

Let $[\mathbf{A}(z)]_{ij} = \alpha_{ij}(z)$, $[\mathbf{C}(z)]_{ik} = \gamma_{ik}(z)$, and $[\mathbf{B}(z)]_{kj} = \beta_{kj}(z)$. Then, the (i,j) element of $\mathbf{A}(z)$ is given by

$$\alpha_{ij}(z) = \sum_{k=1}^{m} \gamma_{ik}(z)\, \beta_{kj}(z).$$

Let $\left[\widetilde{\mathbf{A}}(z)\right]_{ij} = \alpha_{ji}^*(1/z^*)$. Then,

$$\alpha_{ji}^*(1/z^*) = \sum_{k=1}^{m} \beta_{jk}^*(1/z^*)\gamma_{ki}^*(1/z^*) = \left[\widetilde{\mathbf{B}}(z)\widetilde{\mathbf{C}}(z)\right]_{ij}.$$

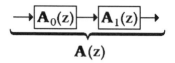

Figure 3.18. Cascaded paraunitary blocks.

Hence, $\widetilde{\mathbf{A}}(z)=\widetilde{\mathbf{B}}(z)\widetilde{\mathbf{C}}(z)$.

■

Now, consider the cascaded paraunitary system of $\mathbf{A}_0(z)$ and $\mathbf{A}_1(z)$ depicted in Figure 3.18. The following theorem provides the intuitively pleasing result that the cascade of two paraunitary systems is a paraunitary system.

Theorem 3.3.2.2. Let $\mathbf{A}(z) = \mathbf{A}_1(z)\mathbf{A}_0(z)$. If $\mathbf{A}_1(z)$ and $\mathbf{A}_0(z)$ are paraunitary, then $\mathbf{A}(z)$ is paraunitary.

Proof: Since $\mathbf{A}_1(z)$ and $\mathbf{A}_0(z)$ are assumed to be paraunitary, then for some $c > 0$ and $d > 0$

$$\widetilde{\mathbf{A}}_1(z)\mathbf{A}_1(z) = d\mathbf{I} \text{ and } \widetilde{\mathbf{A}}_0(z)\mathbf{A}_0(z) = c\mathbf{I} .$$

Hence,

$$\widetilde{\mathbf{A}}(z)\mathbf{A}(z) = \widetilde{\mathbf{A}}_0(z)\widetilde{\mathbf{A}}_1(z)\mathbf{A}_1(z)\mathbf{A}_0(z).$$

Since $\widetilde{\mathbf{A}}_1(z)\mathbf{A}_1(z) = d\mathbf{I}$, then $\widetilde{\mathbf{A}}(z)\mathbf{A}(z)$ becomes

$$\widetilde{\mathbf{A}}(z)\mathbf{A}(z) = d\widetilde{\mathbf{A}}_0(z)\mathbf{A}_0(z).$$

Moreover, $\widetilde{\mathbf{A}}_0(z)\mathbf{A}_0(z) = c\mathbf{I}$. Therefore,

$$\widetilde{\mathbf{A}}(z)\mathbf{A}(z) = dc\mathbf{I}$$

and, as such, $\mathbf{A}(z)$ is a paraunitary matrix.   ■

3.3.2.2 Exploring alias-free filter banks

As we have just observed, alias-free filter banks constrain the matrix $\mathbf{P}(z)$ to be pseudocirculant. Now, using that as a starting point, let us examine the theory of perfect reconstruction filter banks.

First, consider the distortion function, which corresponds to an alias-free filter bank, $i.e.$, (see Section 3.2.2)

$$T(z) = \frac{Y(z)}{X(z)} = \frac{1}{M} \sum_{s=0}^{M-1} z^{-(M-1-s)} \sum_{n=0}^{M-1} z^{-n} P_{s,n}(z^M),$$

or equivalently,

$$T(z) = \frac{1}{M} z^{-(M-1)} \sum_{n=0}^{M-1} \left[\sum_{s=0}^{n} P_{s,n}(z^M) z^{-(n-s)} + \sum_{s=n+1}^{M-1} P_{s,n}(z^M) z^{-(n-s)} \right].$$

For simplicity, represent $\mathbf{P}(z)$ in terms of its zeroth row, that is,

$$P_{k,n}(z) = \begin{cases} P_{0,n-k}(z), & \text{for } k = 0,\ldots,n \\ z^{-1} P_{0,n-k+M}(z), & \text{for } k = n+1,\ldots,M-1. \end{cases}$$

Perform a change of variables in the first summation by letting $k = n - s$ and in the second summation by letting $k = n - s + M$. Then, the distortion function becomes

$$T(z) = \frac{1}{M} z^{-(M-1)} \sum_{n=0}^{M-1} \left[\sum_{k=0}^{n} P_{0,k}(z^M) z^{-k} + \sum_{k=n+1}^{M-1} P_{0,k}(z^M) z^{-k} \right],$$

or equivalently,

$$T(z) = z^{-(M-1)} \sum_{k=0}^{M-1} P_{0,k}(z^M) z^{-k}.$$

Thus, the M-channel filter bank is free from aliasing if and only if the $\mathbf{P}(z)$ is pseudocirculant. Moreover, the distortion function, corresponding to this alias-free filter bank, is given by

$$T(z) = z^{-(M-1)} S(z),$$

where $S(z)$ is the polyphase representation of the zeroth row of $\mathbf{P}(z)$, that is,

$$S(z) = \sum_{k=0}^{M-1} P_{0,k}(z^M) z^{-k},$$

or equivalently,

$$S(z) = [P_{0,0}(z^M), P_{0,1}(z^M), \ldots, P_{0,M-1}(z^M)] \mathbf{e}_M(z)$$

where $\mathbf{e}_M(z)$, the delay chain, is given by

$$\mathbf{e}_M(z) = [1, \ z^{-1}, \ldots, z^{-(M-1)}]^T.$$

Then, the kth row $(k > 0)$ is a right circularly shifted version of $S(z)$, or,

$$z^{-k}S(z) \quad = \quad [z^{-M}P_{0,M-k}(z^M), \ldots, z^{-M}P_{0,M-1}(z^M),$$
$$\ldots, P_{0,M-k-1}(z^M)]\mathbf{e}_M(z).$$

The corresponding system of equations is given by

$$\begin{bmatrix} S(z) \\ z^{-1}S(z) \\ \vdots \\ z^{-(M-1)}S(z) \end{bmatrix} = \mathbf{P}(z^M)\mathbf{e}_M(z).$$

Performing a change of variables by replacing z with zW_M^{-k} yields the following equation

$$\begin{bmatrix} S(zW_M^{-k}) \\ (zW_M^{-k})^{-1}S(zW_M^{-k}) \\ \vdots \\ (zW_M^{-k})^{-(M-1)}S(zW_M^{-k}) \end{bmatrix} = \mathbf{P}(z^M(W_M^{-k})^M)\mathbf{e}_M(zW_M^{-k}),$$

or equivalently,

$$\begin{bmatrix} S(zW_M^{-k}) \\ (zW_M^{-k})^{-1}S(zW_M^{-k}) \\ \vdots \\ (zW_M^{-k})^{-(M-1)}S(zW_M^{-k}) \end{bmatrix} = \mathbf{P}(z^M)\mathbf{e}_M(zW_M^{-k})$$

where $\mathbf{e}_M(zW_M^{-k}) = \left[1, (zW_M^{-k})^{-1}, \ldots, (zW_M^{-k})^{-(M-1)}\right]^T$. This set of equations can be described by

$$\mathbf{\Lambda}(z)\begin{bmatrix} S(zW_M^{-k}) \\ W_M^k S(zW_M^{-k}) \\ \vdots \\ W_M^{(M-1)k}S(zW_M^{-k}) \end{bmatrix} = \mathbf{P}(z^M)\mathbf{\Lambda}(z)\begin{bmatrix} 1 \\ W_M^k \\ \vdots \\ W_M^{(M-1)k} \end{bmatrix},$$

where
$$\Lambda(z) = \text{diag}[1, \; z^{-1}, \ldots, z^{-(M-1)}].$$
Since this equation holds for all values of k for $k = 0, \ldots, M - 1$, we can write it as
$$\Lambda(z)\mathbf{W}_M\mathbf{Q}(z) = \mathbf{P}(z^M)\Lambda(z)\mathbf{W}_M$$
where $\mathbf{W}_M$ is the M x M DFT matrix and
$$\mathbf{Q}(z) = \text{diag} \; [S(z), \ldots, S(zW_M^{-(M-1)})].$$
Thus, any pseudocirculant matrix can be diagonalized.

3.3.2.3 Amplitude distortion-free filter bank

Since $\Lambda(z)$ and $\mathbf{W}_M$ are paraunitary matrices, then $\mathbf{Q}(z)$ is a paraunitary matrix if and only if the pseudocirculant $\mathbf{P}(z)$ is paraunitary. Moreover, $\mathbf{Q}(z)$ is a diagonal matrix with elements $S(zW_M^{-k})$. Therefore, $\mathbf{Q}(z)$ is a paraunitary matrix if and only if $S(z)$ is allpass, that is the magnitude of $S(z)$ is constant, or equivalently, $\mathbf{Q}(z)$ is a paraunitary matrix if and only if $T(z)$ is allpass, since $T(z)$ is simply a delayed version of $S(z)$. Note that $T(z)$ is allpass if and only if the filter bank is amplitude distortion-free with monotonically-varying phase, hence, $T(z)$ is allpass if and only if the pseudocirculant $\mathbf{P}(z)$ is paraunitary or equivalently, the filter bank is amplitude distortion-free with monotonically-varying phase if and only if the pseudocirculant matrix $\mathbf{P}(z)$ is paraunitary.

3.3.2.4 Perfect reconstruction filter banks

Since
$$\mathbf{P}(z^M)\Lambda(z)\mathbf{W}_M = \Lambda(z)\mathbf{W}_M\mathbf{Q}(z)$$
then
$$\det(\mathbf{P}(z^M)) = \det(\mathbf{Q}(z)).$$
However,
$$\det(\mathbf{Q}(z)) = \prod_{k=0}^{M-1} S(zW_M^{-k}).$$
Therefore,
$$\det(\mathbf{P}(z^M)) = \prod_{k=0}^{M-1} S(zW_M^{-k}),$$
or equivalently in the Fourier domain,
$$\det(\mathbf{P}(\exp[j\omega M])) = \prod_{k=0}^{M-1} S\left(\exp\left[j\left(\omega - \frac{2\pi k}{M}\right)\right]\right).$$

Assume $S(z)$ is allpass, then,

$$S\left(\exp\left[j\left(\omega - \frac{2\pi k}{M}\right)\right]\right) = d \exp\left[j\varphi\left(\omega - \frac{2\pi k}{M}\right)\right]$$

for some nonzero d and a monotonically varying phase $\varphi\left(\omega - \frac{2\pi k}{M}\right)$. Hence,

$$\det(\mathbf{P}(\exp[j\omega M])) = d^M \prod_{k=0}^{M-1} \exp\left[j\varphi\left(\omega - \frac{2\pi k}{M}\right)\right],$$

or equivalently,

$$\det(\mathbf{P}(\exp[j\omega M])) = d^M \exp\left[j\sum_{k=0}^{M-1}\varphi\left(\omega - \frac{2\pi k}{M}\right)\right].$$

Assume $S(z)$ has linear phase, then,

$$\varphi(\omega) = g\omega + f$$

for constants g and f. Hence,

$$\sum_{k=0}^{M-1}\varphi\left(\omega - \frac{2\pi k}{M}\right) = \sum_{k=0}^{M-1} g\left(\omega - \frac{2\pi k}{M}\right) + f$$

or equivalently,

$$\sum_{k=0}^{M-1}\varphi\left(\omega - \frac{2\pi k}{M}\right) = M(g\omega + f) - \frac{2\pi g}{M}\sum_{k=1}^{M-1}k.$$

Recall that $\sum_{k=1}^{M-1}k = \frac{M(M-1)}{2}$. Therefore,

$$\sum_{k=0}^{M-1}\varphi\left(\omega - \frac{2\pi k}{M}\right) = Mg\omega + Mf - \pi g(M-1),$$

or equivalently,

$$\sum_{k=0}^{M-1}\varphi\left(\omega - \frac{2\pi k}{M}\right) = a\omega + b$$

where $a = Mg$ and $b = Mf - \pi g(M-1)$. Hence,

$$\det(\mathbf{P}(\exp[j\omega M])) = d^M \exp[j(a\omega + b)]$$

or simply,

$$\det(\mathbf{P}(\exp[j\omega M])) = c \exp(-j\omega m)$$

where $c = d^M \exp(jb)$ and $m = -a$. In the z-transform domain, this equation becomes

$$\det(\mathbf{P}(z^M)) = cz^{-m}.$$

This equation implies that $P_{0,k}(z)$ must have a single nonzero term. Let $\mathbf{p}_0(z)$ be a vector of the polyphase terms in the zeroth row of $\mathbf{P}(z)$, that is,

$$\mathbf{p}_0(z) = [P_{0,0}(z), P_{0,1}(z), \ldots, P_{0,M-1}(z)].$$

Since each $\mathbf{p}_0(z)$ can have only one nonzero component, then successive circular shifts of $\mathbf{p}_0(z)$ will result in M different $\mathbf{P}(z)$ matrices, each of which corresponds to a different perfect reconstruction system. Now, we will enumerate them using the notation $\mathbf{p}_0^{(k)}(z)$ to denote the fact that k circular shifts have been applied to $\mathbf{p}_0(z)$. Now, assume the initial unshifted vector $\mathbf{p}_0^{(0)}(z)$ is given by

$$\mathbf{p}_0^{(0)}(z) = [cz^{-m}, 0, \ldots, 0, 0, 0],$$

which represents the pseudocirculant matrix

$$\mathbf{P}(z) = cz^{-m}\mathbf{I}_M.$$

Apply one circular shift to $\mathbf{p}_0^{(0)}(z)$ to yield $\mathbf{p}_0^{(1)}(z)$, that is,

$$\mathbf{p}_0^{(1)}(z) = [0, cz^{-m}, \ldots, 0, 0, 0].$$

The corresponding pseudocirculant matrix

$$\mathbf{P}(z) = cz^{-m} \begin{bmatrix} \mathbf{0} & \mathbf{I}_{M-1} \\ z^{-1}\mathbf{I}_1 & \mathbf{0} \end{bmatrix}.$$

Therefore, after k circular shifts of $\mathbf{p}_0^{(0)}(z)$, we obtain

$$\mathbf{p}_0^{(k)}(z) = [\underbrace{0, \ldots, 0}_{k \text{ zeros}}, cz^{-M}, \underbrace{0, \ldots, 0}_{M-(k+1) \text{ zeros}}]$$

which represents

$$\mathbf{P}(z) = cz^{-m} \begin{bmatrix} \mathbf{0} & \mathbf{I}_{M-k} \\ z^{-1}\mathbf{I}_k & \mathbf{0} \end{bmatrix} \quad \text{for } k = 0, \ldots, M-1.$$

To illustrate this result, a two-channel filter bank has perfect reconstruction property if and only if

$$\mathbf{P}(z) = cz^{-m}\mathbf{I}_2 \quad \text{or} \quad \mathbf{P}(z) = cz^{-m} \begin{bmatrix} 0 & 1 \\ z^{-1} & 0 \end{bmatrix}.$$

3.3.3   Filter bank formulations in retrospect

In this chapter, we introduced two filter bank formulations: alias-component (AC) and polyphase formulations.

The strategy of the AC formulation is to force the coefficients of the aliased terms to zero by imposing conditions on $A_k(z)$, $k = 0, \ldots, M - 1$, the components of the gain vector.

Attributes of Filter Bank	Characterization of $\mathbf{A}(z)$
Alias-Free	$A_k(z) = 0; \ k = 1, \ldots, M - 1$
Perfect Reconstruction	$A_0(z) = z^{-m_0}$ and $A_k(z) = 0; \ k \neq 0$

As we saw in Section 3.2.2, this approach can lead to difficulties. It requires the inversion of the alias-component matrix and, even if this is successful, there is no guarantee that the resulting analysis filters are stable.

The strategy of the polyphase formulation is to assume the desired result and then to impose conditions on the polyphase matrix $\mathbf{P}(z)$ in order to obtain the correct behavior of the filter bank.

Attributes of Filter Bank	Properties of $\mathbf{P}(z)$
Alias-Free	Pseudocirculant
Alias-Free, Amplitude Distortion-Free, and Monotonically-Varying Phase	Paraunitary, Pseudocirculant
Perfect Reconstruction	Paraunitary, Pseudocirculant Linear Phase

3.4   Filter banks for spectral analysis

We will examine two types of filter banks for spectral analysis, namely the DFT filter bank and the cosine-modulated filter bank. Both of the filter banks discussed in this section have the advantage that all their filters are derived from a single prototype filter. In addition, we will see that the cosine-modulated filter bank can achieve perfect reconstruction.

3.4.1   DFT filter bank

Traditionally, the DFT has been used to analyze the spectrum of a given signal and an efficient implementation using FFT makes DFT a popular choice. Let $h(n)$ be a lowpass FIR prototype filter of length N. Then, the bank of analysis filters, $H_k(z)$, $k = 0, \ldots M - 1$, can be defined as

$$H_k(z) = \sum_{n=0}^{N-1} h(n) W_M^{kn} \, z^{-n}.$$

Assume that the filter length $N = mM$, where M is the number of channels in the filter bank and m is an arbitrary integer constant. Using the Integer Division Theorem, let $n = Mp + q$, then $H_k(z)$ becomes

$$H_k(z) = \sum_{q=0}^{M-1} \sum_{p=0}^{m-1} h(Mp+q) W_M^{k(Mp+q)} z^{-(Mp+q)}.$$

Since $W_M^{k(Mp+q)} = W_M^{kMp} W_M^{kq} = W_M^{kq}$, then the polyphase representation of $H_k(z)$ is given by

$$H_k(z) = \sum_{q=0}^{M-1} \sum_{p=0}^{m-1} h(Mp+q) \, W_M^{kq} \, z^{-(Mp+q)},$$

or equivalently,

$$H_k(z) = \sum_{q=0}^{M-1} G_q(z^M) \, W_M^{kq} \, z^{-q}$$

where,

$$G_q(z) = \sum_{p=0}^{m-1} h(Mp+q) \, z^{-p}.$$

Now expressing the analysis filter bank $H_k(z)$, $k = 0, ..., M - 1$, in matrix form yields

$$\mathbf{h}(z) = \mathbf{W}_M \begin{bmatrix} G_0(z^M) \\ z^{-1} G_1(z^M) \\ \vdots \\ z^{-(M-1)} G_{M-1}(-z^M) \end{bmatrix},$$

where $\mathbf{W}_M$ is a M x M DFT matrix. Let

$$\mathbf{g}(z) = \text{diag } [G_0(z), G_1(z), \ldots, G_{M-1}(z)].$$

Then,

$$\mathbf{h}(z) = \mathbf{W}_M \mathbf{g}(z^M) \mathbf{e}_M(z)$$

where the delay chain $\mathbf{e}_M(z) = \left[1, \ z^{-1}, \ldots, \ z^{-(M-1)}\right]^T$. Since the polyphase decomposition of $\mathbf{h}(z)$ is defined to be

$$\mathbf{h}(z) = \mathbf{E}(z^M) \mathbf{e}_M(z),$$

then, by inspection,

$$\mathbf{E}(z^M) = \mathbf{W}_M \mathbf{g}(z^M).$$

So,
$$\mathbf{E}(z) = \mathbf{W}_M \mathbf{g}(z).$$

Assume that the polyphase-component matrix of the synthesis bank $\mathbf{R}(z) = \widetilde{\mathbf{E}}(z)$. Then,
$$\mathbf{P}(z) = \mathbf{R}(z)\mathbf{E}(z) = \widetilde{\mathbf{E}}(z)\mathbf{E}(z),$$

or equivalently,
$$\mathbf{P}(z) = \widetilde{\mathbf{g}}(z)\mathbf{W}_M^H \mathbf{W}_M \mathbf{g}(z).$$

Since $\mathbf{W}_M^H \mathbf{W}_M = M\mathbf{I}_M$, then
$$\mathbf{P}(z) = M\widetilde{\mathbf{g}}(z)\mathbf{g}(z),$$

or equivalently, $\mathbf{P}(z)$ is a diagonal matrix whose elements are given by
$$[\mathbf{P}(z)]_{k,k} = M\ \widetilde{G}_k(z)G_k(z); \quad k = 0,\ldots,M-1.$$

Since $\mathbf{P}(z)$ is a diagonal matrix, it can become pseudocirculant if all the diagonal elements are equal, that is,
$$[\mathbf{P}(z)]_{0,0} = [\mathbf{P}(z)]_{1,1} = \cdots = [\mathbf{P}(z)]_{M-1,\ M-1}.$$

Once $\mathbf{P}(z)$ is pseudocirculant, then, by definition, the filter bank is alias-free. But, if all the diagonal elements are equal, that is,
$$G_0(z) = G_1(z) = \cdots = G_{M-1}(z),$$

then this corresponds to a square window for the DFT computation, which will cause artifacts in the form of large main and side lobes. To reduce these artifacts we could use overlapping windows, but this would force us to give up the alias-free property. A way around this difficulty is to use cosine-modulated filter banks.

3.4.2   Cosine-modulated filter banks

This study of cosine-modulated filter banks will be broken into two parts. First, we will define reversal matrices and then we will examine properties of cosine-modulated matrices. Then, we will analyze cosine-modulated filter banks.

3.4.2.1   Reversal matrices

Definition 3.4.2.1. *Reversal matrices are matrices with zeroes in all entries except on the antidiagonal. Reversal matrices of dimension m are denoted by $\mathbf{J}_m$.*

For example, $\mathbf{J}_3$ is given by:

$$\mathbf{J}_3 = \begin{bmatrix} 0 & 0 & 1 \\ 0 & 1 & 0 \\ 1 & 0 & 0 \end{bmatrix}.$$

3.4.2.2 Cosine-modulated matrices

In order to analyze cosine-modulated filter banks, we will need to provide a couple of theorems.

Theorem 3.4.2.1. For each $k, q = 0, \ldots, N - 1$, let $c_{k,q} = 2\cos[(2k + 1)\frac{\pi}{2M}(q - mM + \frac{1}{2}) + (-1)^k\frac{\pi}{4}]$. Then, $c_{k,(2Mp+q)} = (-1)^p c_{k,q}$.

Proof: By definition $c_{k,(2Mp+q)}$ is given by

$$c_{k,(2Mp+q)} = 2\cos\left[(2k+1)\frac{\pi}{2M}\left(2Mp + q - mM + \frac{1}{2}\right) + (-1)^k\frac{\pi}{4}\right],$$

or equivalently,

$$\begin{aligned} c_{k,(2Mp+q)} &= 2\cos\left[(2k+1)\frac{\pi}{2M}(2Mp) + (2k+1)\frac{\pi}{2M}\left(q - mM + \frac{1}{2}\right) \right. \\ &\quad \left. + (-1)^k\frac{\pi}{4}\right]. \end{aligned}$$

Recall that for two angles α and β, $\cos(\alpha + \beta) = \cos\alpha\cos\beta - \sin\alpha\sin\beta$. Hence,

$$\begin{aligned} c_{k,(2Mp+q)} &= 2\cos[(2k+1)p\pi]\cos[(2k+1)\frac{\pi}{2M}(q - mM + \frac{1}{2}) \\ &\quad + (-1)^k\frac{\pi}{4}] + 2\sin[(2k+1)p\pi] \\ &\quad \times \sin[(2k+1)\frac{\pi}{2M}(q - mM + \frac{1}{2}) + (-1)^k\frac{\pi}{4}]. \end{aligned}$$

Since $\sin[(2k+1)p\pi] = 0$ and $\cos[(2k+1)p\pi] = (-1)^p$, then

$$c_{k,(2Mp+q)} = (-1)^p c_{k,q}.$$

∎

Theorem 3.4.2.2. $\mathbf{C}'^T\mathbf{C}' = 2M\mathbf{I}_{2M} + 2M(-1)^{(m-1)}\begin{bmatrix} \mathbf{J}_M & 0 \\ 0 & -\mathbf{J}_M \end{bmatrix}$, where $\mathbf{J}_M$ is an M x M reversal matrix and for every $c_{k,q} = \left[\mathbf{C}'\right]_{k,q}$ is described by

$$c_{k,q} = 2\cos[(2k+1)\frac{\pi}{2M}\left(q - mM + \frac{1}{2}\right) + (-1)^k\frac{\pi}{4}].$$

Proof: Let us represent $\mathbf{C}'$ as a partitioned matrix, *i.e.*

$$\mathbf{C}' = \begin{bmatrix} \mathbf{A}_0' & \mathbf{A}_1' \end{bmatrix},$$

where $\mathbf{A}_0'$ and $\mathbf{A}_1'$ are $M \times M$ matrices and $\begin{bmatrix} \mathbf{A}_0' \end{bmatrix}_{k,q} = c_{k,q}$, $\begin{bmatrix} \mathbf{A}_0' \end{bmatrix}_{k,q} = c_{k,q+M}$. Then,

$$\mathbf{C}'^T \mathbf{C}' = \begin{bmatrix} \mathbf{A}_0'^T \\ \mathbf{A}_1'^T \end{bmatrix} \begin{bmatrix} \mathbf{A}_0' & \mathbf{A}_1' \end{bmatrix} = \begin{bmatrix} \mathbf{A}_0'^T \mathbf{A}_0' & \mathbf{A}_0'^T \mathbf{A}_1' \\ \mathbf{A}_1'^T \mathbf{A}_0' & \mathbf{A}_1'^T \mathbf{A}_1' \end{bmatrix}.$$

Thus, proving that $\mathbf{C}'^T \mathbf{C}' = 2M\mathbf{I}_{2M} + 2M(-1)^{(m-1)} \begin{bmatrix} \mathbf{J}_M & 0 \\ 0 & -\mathbf{J}_M \end{bmatrix}$ is equivalent to proving the following relationships:

$$\begin{aligned}
\mathbf{A}_0'^T \mathbf{A}_0' &= 2M[\mathbf{I}_M + (-1)^{(m-1)}\mathbf{J}_M] \\
\mathbf{A}_0'^T \mathbf{A}_1' &= 0 \\
\mathbf{A}_1'^T \mathbf{A}_0' &= 0 \\
\mathbf{A}_1'^T \mathbf{A}_1' &= 2M[\mathbf{I}_M - (-1)^{(m-1)}\mathbf{J}_M].
\end{aligned}$$

We will prove these equations by first examining the $m = 1$ case. Let $\mathbf{A}_0 = \mathbf{A}_0'|_{m=1}$ and $\mathbf{A}_1 = \mathbf{A}_1'|_{m=1}$. Then,

$$[\mathbf{A}_0]_{k,q} = 2\cos\left[(2k+1)\frac{\pi}{2M}\left(q - M + \frac{1}{2}\right) + (-1)^k \frac{\pi}{4}\right]$$

and

$$[\mathbf{A}_1]_{k,q} = 2\cos\left[(2k+1)\frac{\pi}{2M}\left(q + M - M + \frac{1}{2}\right) + (-1)^k \frac{\pi}{4}\right],$$

or equivalently,

$$[\mathbf{A}_1]_{k,q} = 2\cos\left[(2k+1)\frac{\pi}{2M}\left(q + \frac{1}{2}\right) + (-1)^k \frac{\pi}{4}\right].$$

Recall that $\cos(\alpha + \beta) = \cos\alpha\cos\beta - \sin\alpha\sin\beta$. Then, $[\mathbf{A}_0]_{k,q}$ and $[\mathbf{A}_1]_{k,q}$ become

$$\begin{aligned}
[\mathbf{A}_0]_{k,q} = \ & 2\cos[(2k+1)\tfrac{\pi}{2M}(q - M + \tfrac{1}{2})]\cos[(-1)^k \tfrac{\pi}{4}] \\
& -2\sin[(2k+1)\tfrac{\pi}{2M}(q - M + \tfrac{1}{2})]\sin[(-1)^k \tfrac{\pi}{4}]
\end{aligned}$$

and

$$\begin{aligned}
[\mathbf{A}_1]_{k,q} = \ & 2\cos[(2k+1)\tfrac{\pi}{2M}(q + \tfrac{1}{2})]\cos[(-1)^k \tfrac{\pi}{4}] \\
& -2\sin[(2k+1)\tfrac{\pi}{2M}(q + \tfrac{1}{2})]\sin[(-1)^k \tfrac{\pi}{4}].
\end{aligned}$$

Since $\cos[(-1)^k \frac{\pi}{4}] = \frac{\sqrt{2}}{2}$ and $\sin[(-1)^k \frac{\pi}{4}] = (-1)^k \frac{\sqrt{2}}{2}$, then $[\mathbf{A}_0]_{k,q}$ and $[\mathbf{A}_1]_{k,q}$ become

$$[\mathbf{A}_0]_{k,q} = \sqrt{2} \cos[(2k+1)\tfrac{\pi}{2M}(q - M + \tfrac{1}{2})]$$
$$-(-1)^k \sqrt{2} \sin[(2k+1)\tfrac{\pi}{2M}(q - M + \tfrac{1}{2})]$$

and

$$[\mathbf{A}_1]_{k,q} = \sqrt{2} \cos[(2k+1)\tfrac{\pi}{2M}(q + \tfrac{1}{2})]$$
$$-(-1)^k \sqrt{2} \sin[(2k+1)\tfrac{\pi}{2M}(q + \tfrac{1}{2})],$$

or equivalently,

$$[\mathbf{A}_0]_{k,q} = \sqrt{2} \cos[(2k+1)\tfrac{\pi}{2M}(q + \tfrac{1}{2}) - (2k+1)\tfrac{\pi}{2}]$$
$$-(-1)^k \sqrt{2} \sin[(2k+1)\tfrac{\pi}{2M}(q + \tfrac{1}{2}) - (2k+1)\tfrac{\pi}{2}]$$

and

$$[\mathbf{A}_1]_{k,q} = \sqrt{2} \cos[(2k+1)\tfrac{\pi}{2M}(q + \tfrac{1}{2})]$$
$$-(-1)^k \sqrt{2} \sin[(2k+1)\tfrac{\pi}{2M}(q + \tfrac{1}{2})].$$

Recall that $\cos(\alpha - \beta) = \cos\alpha\cos\beta + \sin\alpha\sin\beta$ and $\sin(\alpha - \beta) = \sin\alpha\cos\beta - \cos\alpha\sin\beta$. Then, $[\mathbf{A}_0]_{k,q}$ and $[\mathbf{A}_1]_{k,q}$ become

$$[\mathbf{A}_0]_{k,q} = \sqrt{2} \cos[(2k+1)\tfrac{\pi}{2M}(q + \tfrac{1}{2})] \cos[(2k+1)\tfrac{\pi}{2}]$$
$$+\sqrt{2} \sin[(2k+1)\tfrac{\pi}{2M}(q + \tfrac{1}{2})] \sin[(2k+1)\tfrac{\pi}{2}]$$
$$-(-1)^k \sqrt{2} \sin[(2k+1)\tfrac{\pi}{2M}(q + \tfrac{1}{2})] \cos[(2k+1)\tfrac{\pi}{2}]$$
$$+(-1)^k \sqrt{2} \cos[(2k+1)\tfrac{\pi}{2M}(q + \tfrac{1}{2})] \sin[(2k+1)\tfrac{\pi}{2}]$$

and

$$[\mathbf{A}_1]_{k,q} = \sqrt{2} \cos[(2k+1)\tfrac{\pi}{2M}(q + \tfrac{1}{2})]$$
$$-(-1)^k \sqrt{2} \sin[(2k+1)\tfrac{\pi}{2M}(q + \tfrac{1}{2})].$$

Since $\cos[(2k+1)\tfrac{\pi}{2}] = 0$ and $\sin[(2k+1)\tfrac{\pi}{2}] = (-1)^k$, then $[\mathbf{A}_0]_{k,q}$ and $[\mathbf{A}_1]_{k,q}$ become

$$[\mathbf{A}_0]_{k,q} = \sqrt{2} \cos[(2k+1)\tfrac{\pi}{2M}(q + \tfrac{1}{2})]$$
$$+(-1)^k \sqrt{2} \sin[(2k+1)\tfrac{\pi}{2M}(q + \tfrac{1}{2})]$$

and

$$[\mathbf{A}_1]_{k,q} = \sqrt{2} \cos[(2k+1)\tfrac{\pi}{2M}(q + \tfrac{1}{2})]$$
$$-(-1)^k \sqrt{2} \sin[(2k+1)\tfrac{\pi}{2M}(q + \tfrac{1}{2})].$$

Let $\mathcal{C}$ and $\mathcal{S}$ denote Type-IV discrete cosine transforms (DCT) and Type-IV discrete sine transforms (DST) matrices, respectively. Then, the elements of $\mathcal{C}$ and $\mathcal{S}$ are given by

$$[\mathcal{C}]_{kq} = \sqrt{\frac{2}{M}} \cos\left[(2k+1)\frac{\pi}{2M}\left(q+\frac{1}{2}\right)\right]$$

and

$$[\mathcal{S}]_{kq} = \sqrt{\frac{2}{M}} \sin\left[(2k+1)\frac{\pi}{2M}\left(q+\frac{1}{2}\right)\right].$$

Let Λ be an M x M matrix defined by

$$[\Lambda]_{kq} = (-1)^k \delta_{k,q}; \quad k, q = 0, \ldots, M-1.$$

Then,

$$[\Lambda\mathcal{S}]_{kq} = (-1)^k [\mathcal{S}]_{kq}.$$

Therefore, $[\mathbf{A}_0]_{k,q}$ and $[\mathbf{A}_1]_{k,q}$ can be written as

$$[\mathbf{A}_0]_{k,q} = \sqrt{M} [\mathcal{C} + \Lambda\mathcal{S}]_{kq}$$

and

$$[\mathbf{A}_1]_{k,q} = \sqrt{M} [\mathcal{C} - \Lambda\mathcal{S}]_{kq}.$$

Consider the equations for $\mathbf{A}_0^T\mathbf{A}_0$, $\mathbf{A}_0^T\mathbf{A}_1$, $\mathbf{A}_1^T\mathbf{A}_0$, and $\mathbf{A}_1^T\mathbf{A}_1$.

$$
\begin{aligned}
\mathbf{A}_0^T\mathbf{A}_0 &= M(\mathcal{C} + \Lambda\mathcal{S})^T(\mathcal{C} + \Lambda\mathcal{S}) \\
&= M\left(\mathcal{C}^T\mathcal{C} + \mathcal{C}^T\Lambda\mathcal{S} + \mathcal{S}^T\Lambda^T\mathcal{C} + \mathcal{S}^T\Lambda^T\Lambda\mathcal{S}\right),
\end{aligned}
$$

$$
\begin{aligned}
\mathbf{A}_0^T\mathbf{A}_1 &= M(\mathcal{C} + \Lambda\mathcal{S})^T(\mathcal{C} - \Lambda\mathcal{S}) \\
&= M\left(\mathcal{C}^T\mathcal{C} - \mathcal{C}^T\Lambda\mathcal{S} + \mathcal{S}^T\Lambda^T\mathcal{C} - \mathcal{S}^T\Lambda^T\Lambda\mathcal{S}\right),
\end{aligned}
$$

$$
\begin{aligned}
\mathbf{A}_1^T\mathbf{A}_0 &= M(\mathcal{C} - \Lambda\mathcal{S})^T(\mathcal{C} + \Lambda\mathcal{S}) \\
&= M\left(\mathcal{C}^T\mathcal{C} + \mathcal{C}^T\Lambda\mathcal{S} - \mathcal{S}^T\Lambda^T\mathcal{C} - \mathcal{S}^T\Lambda^T\Lambda\mathcal{S}\right),
\end{aligned}
$$

and

$$
\begin{aligned}
\mathbf{A}_1^T\mathbf{A}_1 &= M(\mathcal{C} - \Lambda\mathcal{S})^T(\mathcal{C} - \Lambda\mathcal{S}) \\
&= M\left(\mathcal{C}^T\mathcal{C} - \mathcal{C}^T\Lambda\mathcal{S} - \mathcal{S}^T\Lambda^T\mathcal{C} + \mathcal{S}^T\Lambda^T\Lambda\mathcal{S}\right).
\end{aligned}
$$

But, $[\Lambda^T\Lambda]_{k,k} = (-1)^k(-1)^k = 1$. Therefore, $\Lambda^T\Lambda = \mathbf{I}_M$. In addition, since the Type-IV DCT and DST matrices are orthogonal, then $\mathcal{C}^T\mathcal{C} = \mathbf{I}_M$ and $\mathcal{S}^T\mathcal{S} = \mathbf{I}_M$. Hence,

$$
\begin{aligned}
\mathbf{A}_0^T\mathbf{A}_0 &= M\left(2\mathbf{I}_M + \mathcal{C}^T\Lambda\mathcal{S} + \mathcal{S}^T\Lambda^T\mathcal{C}\right) \\
\mathbf{A}_0^T\mathbf{A}_1 &= M\left(-\mathcal{C}^T\Lambda\mathcal{S} + \mathcal{S}^T\Lambda^T\mathcal{C}\right) \\
\mathbf{A}_1^T\mathbf{A}_0 &= M\left(\mathcal{C}^T\Lambda\mathcal{S} - \mathcal{S}^T\Lambda^T\mathcal{C}\right) \\
\mathbf{A}_1^T\mathbf{A}_1 &= M\left(2\mathbf{I}_M - \mathcal{C}^T\Lambda\mathcal{S} - \mathcal{S}^T\Lambda^T\mathcal{C}\right).
\end{aligned}
$$

Let $\mathbf{J}_M$ be the M x M reversal matrix. Then, $\Lambda \mathcal{S} \mathbf{J}_M = \mathcal{C}$. Moreover, since $\mathbf{J}_M^2 = \mathbf{I}_M$, then $\mathcal{C}^T \Lambda \mathcal{S} = \mathbf{J}_M$ and $\mathcal{S}^T \Lambda^T \mathcal{C} = \mathbf{J}_M$. Therefore,

$$
\begin{aligned}
\mathbf{A}_0^T \mathbf{A}_0 &= 2M \left(\mathbf{I}_M + \mathbf{J}_M \right) \\
\mathbf{A}_0^T \mathbf{A}_1 &= 0 \\
\mathbf{A}_1^T \mathbf{A}_0 &= 0 \\
\mathbf{A}_1^T \mathbf{A}_1 &= 2M \left(\mathbf{I}_M - \mathbf{J}_M \right)
\end{aligned}
$$

which proves the special case of $m = 1$. We must now extend our results for arbitrary integer $m > 1$, by expressing the interrelationships between $\{\mathbf{A}_0, \mathbf{A}_1\}$ and $\{\mathbf{A}_0', \mathbf{A}_1'\}$.

$$
\begin{aligned}
\text{For } m \text{ even,} \quad \mathbf{A}_0' &= (-1)^{\frac{m}{2}} \mathbf{A}_1 \\
\mathbf{A}_1' &= (-1)^{\frac{m}{2}-1} \mathbf{A}_0 \\
\text{For } m \text{ odd,} \quad \mathbf{A}_0' &= (-1)^{\frac{m-1}{2}} \mathbf{A}_0 \\
\mathbf{A}_1' &= (-1)^{\frac{m-1}{2}} \mathbf{A}_1.
\end{aligned}
$$

Utilizing these relationships gives the following:

$$
\begin{aligned}
\text{For } m \text{ even,} \quad \mathbf{A}_0'^T \mathbf{A}_0' &= 2M \left(\mathbf{I}_M - \mathbf{J}_M \right) \\
\mathbf{A}_0'^T \mathbf{A}_1' &= 0 \\
\mathbf{A}_1'^T \mathbf{A}_0' &= 0 \\
\mathbf{A}_1'^T \mathbf{A}_1' &= 2M \left(\mathbf{I}_M + \mathbf{J}_M \right). \\
\text{For } m \text{ odd,} \quad \mathbf{A}_0'^T \mathbf{A}_0' &= 2M \left(\mathbf{I}_M + \mathbf{J}_M \right) \\
\mathbf{A}_0'^T \mathbf{A}_1' &= 0 \\
\mathbf{A}_1'^T \mathbf{A}_0' &= 0 \\
\mathbf{A}_1'^T \mathbf{A}_1' &= 2M \left(\mathbf{I}_M - \mathbf{J}_M \right).
\end{aligned}
$$

Combining the odd and even results yields

$$
\begin{aligned}
\mathbf{A}_0'^T \mathbf{A}_0' &= 2M[\mathbf{I}_M + (-1)^{(m-1)} \mathbf{J}_M] \\
\mathbf{A}_0'^T \mathbf{A}_1' &= 0 \\
\mathbf{A}_1'^T \mathbf{A}_0' &= 0 \\
\mathbf{A}_1'^T \mathbf{A}_1' &= 2M[\mathbf{I}_M - (-1)^{(m-1)} \mathbf{J}_M], \text{or}
\end{aligned}
$$

$$
\mathbf{C}'^T \mathbf{C}' = 2M \mathbf{I}_{2M} + 2M(-1)^{(m-1)} \begin{bmatrix} \mathbf{J}_M & 0 \\ 0 & -\mathbf{J}_M \end{bmatrix}.
$$

∎

3.4.2.3   Cosine-modulated filter banks

Let $h(n)$ be a low-pass FIR prototype filter of length N. Then, the bank
of analysis filters, $H_k(z)$, $k = 0, \ldots M - 1$, can be defined as

$$H_k(z) = \sum_{n=0}^{N-1} c_{k,n} h(n) \, z^{-n},$$

where

$$c_{k,n} = 2 \cos \left[(2k + 1) \frac{\pi}{2M} \left(n - mM + \frac{1}{2} \right) + (-1)^k \frac{\pi}{4} \right]$$

and $h(n)$ is a linear phase, low-pass FIR prototype filter. Assume that the
filter length $N = m2M$, where M is the number of channels in the filter
bank and m is an arbitrary integer constant. Using the Integer Division
Theorem, let $n = (2M)p + q$, then $H_k(z)$ becomes

$$H_k(z) = \sum_{q=0}^{2M-1} \sum_{p=0}^{m-1} c_{k,(2Mp+q)} \, h(2Mp + q) \, z^{-(2Mp+q)}.$$

Since $c_{k,(2Mp+q)} = (-1)^p c_{k,q}$, then the polyphase representation of $H_k(z)$
is given by

$$H_k(z) = \sum_{q=0}^{2M-1} c_{k,q} z^{-q} \sum_{p=0}^{m-1} (-1)^p \, h(2Mp + q) \, z^{-2Mp},$$

or equivalently,

$$H_k(z) = \sum_{q=0}^{2M-1} c_{k,q} G_q(-z^{2M}) z^{-q}$$

where,

$$G_q(z) = \sum_{p=0}^{m-1} h(q + 2pM) \, z^{-p}.$$

Now, expressing the analysis filter bank $H_k(z)$, $k = 0, \ldots, M - 1$, in matrix
form yields

$$\mathbf{h}(z) = \mathbf{C}' \begin{bmatrix} G_0(-z^{2M}) \\ z^{-1} G_1(-z^{2M}) \\ \vdots \\ z^{-(2M-1)} G_{2M-1}(-z^{2M}) \end{bmatrix},$$

where $\mathbf{C}'$ is a $M \times 2M$ matrix defined by

$$\left[\mathbf{C}' \right]_{k,n} = 2 \cos[(2k + 1) \frac{\pi}{2M} (n - \frac{N-1}{2}) + (-1)^k \frac{\pi}{4}].$$

Let

$$\mathbf{g}_0(-z) = \text{diag} \ [G_0(-z), G_1(-z), \ldots, G_{M-1}(-z)]$$

and

$$\mathbf{g}_1(-z) = \text{diag} \ [G_M(-z), G_{M+1}(-z), \ldots, G_{2M-1}(-z)].$$

Then,

$$\mathbf{h}(z) = \mathbf{C}' \left[\begin{array}{cc} \mathbf{g}_0(-z^{2M}) & 0 \\ 0 & \mathbf{g}_1(-z^{2M}) \end{array} \right] \left[\begin{array}{c} \mathbf{e}_M(z) \\ z^{-M}\mathbf{e}_M(z) \end{array} \right],$$

or equivalently,

$$\mathbf{h}(z) = \mathbf{C}' \left[\begin{array}{c} \mathbf{g}_0(-z^{2M}) \\ z^{-M}\mathbf{g}_1(-z^{2M}) \end{array} \right] \mathbf{e}_M(z),$$

where the delay chain $\mathbf{e}_M(z) = \left[1, \ z^{-1}, \ldots, \ z^{-(M-1)}\right]^T$. Since the polyphase decomposition of $\mathbf{h}(z)$ is defined to be

$$\mathbf{h}(z) = \mathbf{E}(z^M)\mathbf{e}_M(z),$$

then, by inspection,

$$\mathbf{E}(z^M) = \mathbf{C}' \left[\begin{array}{c} \mathbf{g}_0(-z^{2M}) \\ z^{-M}\mathbf{g}_1(-z^{2M}) \end{array} \right].$$

So,

$$\mathbf{E}(z) = \mathbf{C}' \left[\begin{array}{c} \mathbf{g}_0(-z^2) \\ z^{-1}\mathbf{g}_1(-z^2) \end{array} \right].$$

Assume that the polyphase-component matrix of the synthesis bank $\mathbf{R}(z) = \widetilde{\mathbf{E}}(z)$. Then,

$$\mathbf{P}(z) = \widetilde{\mathbf{E}}(z)\mathbf{E}(z)$$

or equivalently,

$$\mathbf{P}(z) = \left[\begin{array}{cc} \widetilde{\mathbf{g}}_0(-z^2) & z\widetilde{\mathbf{g}}_1(-z^2) \end{array} \right] \mathbf{C}'^T \mathbf{C}' \left[\begin{array}{c} \mathbf{g}_0(-z^2) \\ z^{-1}\mathbf{g}_1(-z^2) \end{array} \right].$$

Since $\mathbf{C}'^T\mathbf{C}' = 2M\mathbf{I}_{2M} + 2M(-1)^{(m-1)} \left[\begin{array}{cc} \mathbf{J}_M & 0 \\ 0 & -\mathbf{J}_M \end{array} \right]$, then

$$\begin{aligned} \mathbf{P}(z) \ = \ & 2M \left(\widetilde{\mathbf{g}}_0(-z^2)\mathbf{g}_0(-z^2) + \widetilde{\mathbf{g}}_1(-z^2)\mathbf{g}_1(-z^2)\right) \\ & + 2M(-1)^{(m-1)} \left(\widetilde{\mathbf{g}}_0(-z^2)\mathbf{J}_M\mathbf{g}_0(-z^2) - \widetilde{\mathbf{g}}_1(-z^2)\mathbf{J}_M\mathbf{g}_1(-z^2)\right). \end{aligned}$$

Assume $\mathbf{H}(z)$ has linear phase. Hence, the polyphase-components of $\mathbf{H}(z)$ are related by

$$G_k(z) = z^{-(m-1)} \widetilde{G}_{2M-1-k}(z).$$

Since

$$\mathbf{g}_0(-z^2) = \text{diag}\left[G_{2M-1}(-z^2), G_{2M-2}(-z^2), \ldots, G_M(-z^2)\right]$$

and

$$\mathbf{g}_1(-z^2) = \text{diag}\left[G_{M-1}(-z^2), G_{M-2}(-z^2), \ldots, G_0(-z^2)\right]$$

then

$$\widetilde{\mathbf{g}}_0(-z^2) = z^{-2(m-1)} \mathbf{J}_M \text{ diag}\left[\widetilde{G}_M(-z^2), \widetilde{G}_{M+1}(-z^2), \ldots, \widetilde{G}_{2M-1}(-z^2)\right]$$

and

$$\widetilde{\mathbf{g}}_1(-z^2) = z^{-2(m-1)} \mathbf{J}_M \text{ diag}\left[\widetilde{G}_0(-z^2), \widetilde{G}_1(-z^2), \ldots, \widetilde{G}_{M-1}(-z^2)\right]$$

or equivalently,

$$\widetilde{\mathbf{g}}_0(-z^2) = z^{-2(m-1)} \mathbf{J}_M \widetilde{\mathbf{g}}_1(-z^2)$$

and

$$\widetilde{\mathbf{g}}_1(-z^2) = z^{-2(m-1)} \mathbf{J}_M \widetilde{\mathbf{g}}_0(-z^2).$$

Substituting these equations for $\widetilde{\mathbf{g}}_0(-z^2)$ and $\widetilde{\mathbf{g}}_1(-z^2)$ into the equation for $\mathbf{P}(z)$ yields

$$
\begin{aligned}
\mathbf{P}(z) =\;& 2M\left(\widetilde{\mathbf{g}}_0(-z^2)\mathbf{g}_0(-z^2) + \widetilde{\mathbf{g}}_1(-z^2)\mathbf{g}_1(-z^2)\right) \\
&+ 2M(-1)^{(m-1)} z^{2(m-1)}\left(\widetilde{\mathbf{g}}_0(-z^2)\widetilde{\mathbf{g}}_1(-z^2) - \widetilde{\mathbf{g}}_1(-z^2)\mathbf{g}_0(-z^2)\right).
\end{aligned}
$$

Since $\widetilde{\mathbf{g}}_0(-z^2)$ and $\widetilde{\mathbf{g}}_1(-z^2)$ are both diagonal matrices, then $\widetilde{\mathbf{g}}_0(-z^2)$ commutes with $\widetilde{\mathbf{g}}_1(-z^2)$, i.e.

$$\widetilde{\mathbf{g}}_0(-z^2)\widetilde{\mathbf{g}}_1(-z^2) = \widetilde{\mathbf{g}}_1(-z^2)\mathbf{g}_0(-z^2).$$

Therefore, the second term of the equation for $\mathbf{P}(z)$ equals zero. Hence,

$$\mathbf{P}(z) = 2M\left(\widetilde{\mathbf{g}}_0(-z^2)\mathbf{g}_0(-z^2) + \widetilde{\mathbf{g}}_1(-z^2)\mathbf{g}_1(-z^2)\right),$$

or equivalently, $\mathbf{P}(z)$ is a diagonal matrix whose elements are given by

$$[\mathbf{P}(z)]_{k,k} = 2M\left(\widetilde{G}_k(-z)G_k(-z) + \widetilde{G}_{M+k}(-z)G_{M+k}(-z)\right).$$

Since $\mathbf{P}(z)$ is a diagonal matrix, it can become pseudocirculant if all the diagonal elements are equal, i.e.,

$$[\mathbf{P}(z)]_{0,0} = [\mathbf{P}(z)]_{1,1} = \cdots = [\mathbf{P}(z)]_{M-1,\ M-1}.$$

Once $\mathbf{P}(z)$ is pseudocirculant, then, by definition, the filter bank is alias-free. Moreover, if all the diagonal elements of pseudocirculant $\mathbf{P}(z)$ are equal to a constant times a delay, then the cosine-modulated filter bank satisfies the perfect reconstruction property.

3.4.3 A generalization: Malvar wavelets

Let us begin by partitioning the real line $\mathbf{R}$. Let $\mathbf{Z}$ denote the integers and let $\mathbf{N}$ denote the natural numbers. Let $\{a_j| \ j \in \mathbf{Z}\}$ denote a strictly increasing sequence such that $a_j \rightarrow -\infty$ as $j \rightarrow -\infty$ and $a_j \rightarrow +\infty$ as $j \rightarrow +\infty$. The interval $(a_j, a_{j+1}]$ denotes the jth subinterval and $\mathbf{R} = \cup_{j \in \mathbf{Z}}(a_j, a_{j+1}]$ so that the collection $\{(a_j, a_{j+1}]| \ j \in \mathbf{Z}\}$ forms a partition of $\mathbf{R}$. For each $j \in \mathbf{Z}$ let $\{f_{j,k}| \ k \in \mathbf{N}\}$ denote a real-valued orthonormal basis defined on the interval $I_j = [a_j, a_{j+1}]$ where orthogonality is given by

$$\int_{a_j}^{a_{j+1}} f_{j,k}(x) \ f_{j,l}(x)dx = \delta_{k,l}.$$

At each point a_j we center an interval, namely $(a_j - \epsilon_j, a_j + \epsilon_j)$ with $\epsilon_j > 0$ such that $\epsilon_{j+1} + \epsilon_j \leq a_{j+1} - a_j$. We define extension functions $\widehat{f}_{j,k}$ by constructing the odd extension about a_j of $f_{j,k}$ on $(a_j - \epsilon_j, a_j)$ and the even extension about a_{j+1} of $f_{j,k}$ on $(a_{j+1}, a_{j+1} + \epsilon_{j+1})$ in the following way.

$$\widehat{f}_{j,k} = \begin{cases} 0, & -\infty < x \leq a_j - \epsilon_j \\ -f_{j,k}(2a_j - x), & a_j - \epsilon_j < x < a_j \\ f_{j,k}(x), & a_j \leq x \leq a_{j+1} \\ f_{j,k}(2a_{j+1} - x), & a_{j+1} < x < a_{j+1} + \epsilon_{j+1} \\ 0, & a_{j+1} + \epsilon_{j+1} \leq x < \infty. \end{cases}$$

Let ω_j denote a window function which is positive on the interval $(a_j + \epsilon_j, a_{j+1} - \epsilon_{j+1})$ having a peak amplitude of unity. It is important to note that the intersection of the positive support of two nonadjacent windows is the null set. We will choose windows $\omega_j(x)$ with the following properties.

$(a) \quad \omega_j(x) = 1 \qquad \qquad \text{for } x \in (a_j + \epsilon_j, a_{j+1} - \epsilon_{j+1})$

$(b) \quad \omega_j(x) = 0 \qquad \qquad \text{for } x \notin (a_j - \epsilon_j, a_{j+1} + \epsilon_{j+1})$

$(c) \quad \omega_j(x - \sigma) = \omega_{j-1}(x + \sigma) \quad \text{for } \sigma \in [-\epsilon_j, \epsilon_j]$

$(d) \quad \omega_j^2(x) + \omega_{j-1}^2(x) = 1 \qquad \text{for } x \in [a_j - \epsilon_j, a_j + \epsilon_j].$

Definition 3.4.3.1. *We now define Malvar wavelets by the functions $u_{j,k}$ for each $j \in \mathbf{Z}$ and $k \in \mathbf{N}$ by $u_{j,k}(x) = \omega_j(x)\widehat{f}_{j,k}(x)$.*

The following theorem states a very useful property for Malvar wavelets.

Theorem 3.4.3.1. *Malvar wavelets, $\{ u_{j,k}| \ j \in \mathbf{Z}, \ k \in \mathbf{N}\}$, are an orthonormal basis for $L^2(\mathbf{R})$.*

Proof: Observe that $u_{j,k}$ has compact support and by construction satisfies $\|u_{j,k}\|_{L^2(\mathbf{R})} < \infty$ for each $j \in \mathbf{Z}$ and $k \in \mathbf{N}$. Therefore, $\{ u_{j,k} | \, j \in \mathbf{Z}, k \in \mathbf{N}\} \subset L^2(\mathbf{R})$.

Next we demonstrate that $\{ u_{j,k} | \, j \in \mathbf{Z}, \, k \in \mathbf{N}\}$ is an orthonormal set. By the nonoverlapping property of the positive support intervals, it is clear that $\langle u_{j,k}, u_{i,l} \rangle = 0$ if $|i - j| \geq 2$ for all $k, l \in \mathbf{N}$. Therefore, we need to consider two cases: (I) $i = j$ and (II) $|j - i| = 1$.

<u>Case I</u>: We will show $\langle u_{j,k}, u_{j,l} \rangle = \delta_{k,l}$. To see this, observe that

$$\langle u_{j,k}, u_{j,l} \rangle = \int_{a_j - \epsilon_j}^{a_{j+1} + \epsilon_{j+1}} \omega_j^2(x) \widehat{f}_{j,k}(x) \widehat{f}_{j,l}(x) dx.$$

From the definitions of even and odd extensions one sees that

$$
\begin{aligned}
\langle u_{j,k}, u_{j,l} \rangle &= \int_{a_j - \epsilon_j}^{a_j} \omega_j^2(x) f_{j,k}(2a_j - x) f_{j,l}(2a_j - x) dx \\
&\quad + \int_{a_j}^{a_j + \epsilon_j} \left[\omega_j^2(x) - 1 \right] f_{j,k}(x) f_{j,l}(x) dx \\
&\quad + \int_{a_j}^{a_{j+1}} f_{j,k}(x) f_{j,l}(x) dx \\
&\quad + \int_{a_{j+1} - \epsilon_{j+1}}^{a_{j+1}} \left[\omega_j^2(x) - 1 \right] f_{j,k}(x) f_{j,l}(x) dx \\
&\quad + \int_{a_{j+1}}^{a_{j+1} + \epsilon_{j+1}} \omega_j^2(x) f_{j,k}(2a_{j+1} - x) f_{j,l}(2a_{j+1} - x) dx \\
&= A + B + C + D + E.
\end{aligned}
$$

Note that $C = \delta_{k,l}$ by choice. For A, let $x = a_j - \sigma$ and for B, let $x = a_j + \sigma$ to yield

$$A + B = \int_0^{\epsilon_j} \left[\omega_j^2(a_j - \sigma) + \omega_j^2(a_j + \sigma) - 1 \right] f_{j,k}(a_j + \sigma) f_{j,l}(a_j + \sigma) d\sigma.$$

By property (c), $\omega_j(x - \sigma) = \omega_{j-1}(x + \sigma)$ and so property (d) implies

$$\omega_j^2(a_j - \sigma) + \omega_j^2(a_j + \sigma) - 1 = 0 \text{ for } -\epsilon_j \leq \sigma \leq \epsilon_j$$

hence, $A + B = 0$. Similarly, in integral D, let $x = a_{j+1} - \sigma$ and in E, let $x = a_{j+1} + \sigma$ to get

$$
\begin{aligned}
D + E &= \int_0^{\epsilon_{j+1}} \left[\omega_j^2(a_{j+1} - \sigma) + \omega_j^2(a_{j+1} + \sigma) - 1 \right] \\
&\quad \times f_{j,k}(a_{j+1} - \sigma) f_{j,l}(a_{j+1} - \sigma) d\sigma.
\end{aligned}
$$

Properties (c) and (d) imply $D + E = 0$. Consequently,

$$\langle u_{j,k}, u_{j,l} \rangle = \delta_{k,l}.$$

<u>Case II</u>: We will show $\langle u_{i,k}, u_{j,l} \rangle = 0$ for $i = j - 1$ for every $j \in \mathbf{Z}$, $k, l \in \mathbf{N}$ then the case $i = j + 1$ will follow easily.

By construction, $u_{j-1,k}$ and $u_{j,l}$ are possibly nonzero only on $(a_j - \epsilon_j, a_j + \epsilon_j)$, hence

$$\langle u_{j-1,k}, u_{j,l} \rangle = \int_{a_j - \epsilon_j}^{a_j + \epsilon_j} \omega_{j-1}(x)\omega_j(x)\widehat{f}_{j-1,k}(x)\widehat{f}_{j,l}(x)dx.$$

From the definitions of the extensions

$$
\begin{aligned}
\langle u_{j-1,k}, u_{j,l} \rangle &= -\int_{a_j - \epsilon_j}^{a_j} \omega_{j-1}(x)\omega_j(x)f_{j-1,k}(x)f_{j,l}(2a_j - x)dx \\
&\quad + \int_{a_j}^{a_j + \epsilon_j} \omega_{j-1}(x)\omega_j(x)f_{j-1,k}(2a_j - x)f_{j,l}(x)dx.
\end{aligned}
$$

In the first integral, let $x = a_j - \sigma$ and in the second integral, let $x = a_j + \sigma$, then

$$
\begin{aligned}
\langle u_{j-1,k}, u_{j,l} \rangle &= \int_0^{\epsilon_j} [-\omega_{j-1}(a_j - \sigma)\omega_j(a_j - \sigma) + \omega_{j-1}(a_j + \sigma) \\
&\quad \times \omega_j(a_j + \sigma)]f_{j-1,k}(a_j - \sigma)f_{j,l}(a_j + \sigma)d\sigma.
\end{aligned}
$$

Property (c) implies

$$-\omega_{j-1}(a_j - \sigma)\omega_j(a_j - \sigma) + \omega_{j-1}(a_j + \sigma)\omega_j(a_j + \sigma) = 0.$$

Therefore, $\langle u_{j-1,k}, u_{j,l} \rangle = 0$. Cases I and II demonstrate that $\{ u_{j,k} | j \in \mathbf{Z}, k \in \mathbf{N} \}$ is an orthonormal set.

Lastly, we prove that $\{ u_{j,k} | j \in \mathbf{Z}, k \in \mathbf{N} \}$ is a basis for $L^2(\mathbf{R})$. To do this, we will show that given $s \in L^2(\mathbf{R})$ there exists a set of scalars $\{ \alpha_{j,k} | j \in \mathbf{Z}, k \in \mathbf{N} \}$ such that

$$s = \sum_{j \in \mathbf{Z}} \sum_{k \in \mathbf{N}} \alpha_{j,k} u_{j,k}$$

in the $L^2(\mathbf{R})$ sense. Let $s \in L^2(\mathbf{R})$ and define $s_j(x) = s(x)\omega_j(x)$ for each $j \in \mathbf{Z}$. Since s_j has positive support on $(a_j - \epsilon_j, a_{j+1} + \epsilon_{j+1})$ we *fold* $s_j(x)$ on $(a_j - \epsilon_j, a_j)$ and $(a_{j+1}, a_{j+1} + \epsilon_{j+1})$ into the interval $[a_j, a_{j+1}]$ by defining $h_j(x)$ as

$$
h_j(x) = \begin{cases}
0, & -\infty < x < a_j \\
s_j(x) - s_j(2a_j - x), & a_j \leq x \leq a_j + \epsilon_j \\
s_j(x), & a_j + \epsilon_j < x < a_{j+1} - \epsilon_{j+1} \\
s_j(x) + s_j(2a_{j+1} - x), & a_{j+1} - \epsilon_{j+1} \leq x \leq a_{j+1} \\
0, & a_{j+1} < x < \infty.
\end{cases}
$$

Now h_j is supported on $[a_j, a_{j+1}]$. Consequently, there exist real numbers $\alpha_{j,k}$ for $k \in \mathbf{N}$ such that

$$h_j(x) = \sum_{k=1}^{\infty} \alpha_{j,k} f_{j,k}(x)$$

where convergence and equality is in the $L^2(\mathbf{I}_j)$ sense, and $\alpha_{j,k}$ is given by the inner product rule

$$
\begin{aligned}
\alpha_{j,k} &= \langle h_j, f_{j,k} \rangle_{L^2(\mathbf{I}_j)} \\
&= \int_{a_j}^{a_{j+1}} h_j(x) f_{j,k}(x) dx \\
&= \int_{a_j+\epsilon_j}^{a_{j+1}-\epsilon_{j+1}} s_j(x) \widehat{f}_{j,k}(x) dx \\
&= \int_{a_j+\epsilon_j}^{a_{j+1}-\epsilon_{j+1}} s(x) u_{j,k}(x) dx \\
&= \langle s, u_{j,k} \rangle.
\end{aligned}
$$

Applying the function rules at both endpoints yields

$$
\widehat{h}_j(x) = \sum_{k=1}^{\infty} \alpha_{j,k} \widehat{f}_{j,k}(x)
$$

where $\widehat{h}_j$ has an odd extension about $x = a_j$ and an even extension about $x = a_{j+1}$. That is, the use of the extension rules for $\widehat{f}_{j,k}(x)$ applied to $h_j(x)$ yields the following

$$
\widehat{h}_j(x) = \begin{cases}
0, & -\infty < x < a_j - \epsilon_j \\
s_j(x) - s_j(2a_j - x), & a_j - \epsilon_j \le x \le a_j + \epsilon_j \\
s_j(x), & a_j + \epsilon_j < x < a_{j+1} - \epsilon_{j+1} \\
s_j(x) + s_j(2a_{j+1} - x), & a_{j+1} - \epsilon_{j+1} \le x \le a_{j+1} + \epsilon_{j+1} \\
0, & a_{j+1} + \epsilon_{j+1} < x < \infty.
\end{cases}
$$

Multiplying $\widehat{h}_j(x)$ by $\omega_j(x)$ and summing over j produces

$$
\sum_{j \in \mathbf{Z}} \omega_j(x) \widehat{h}_j(x) = \sum_{j \in \mathbf{Z}} \sum_{k \in \mathbf{N}} \alpha_{j,k} \omega_j(x) \widehat{f}_{j,k}(x) = \sum_{j \in \mathbf{Z}} \sum_{k \in \mathbf{N}} \alpha_{j,k} u_{j,k}.
$$

To complete the proof, we will show

$$
\sum_{j \in \mathbf{Z}} \sum_{k \in \mathbf{N}} \alpha_{j,k} u_{j,k}(x) = s(x).
$$

If x is a point where $s(x)$ is defined and $x \in [a_J + \epsilon_J, a_{J+1} + \epsilon_{J+1}]$ for some fixed $J \in \mathbf{Z}$ then

$$
\sum_{j \in \mathbf{Z}} \omega_j(x) \widehat{h}_j(x) = \omega_J(x) h_J(x) = s(x).
$$

If $x \in (a_J - \epsilon_J, a_J + \epsilon_J)$, then

$$
\sum_{j \in \mathbf{Z}} \omega_j(x) \widehat{h}_j(x) = \omega_{J-1}(x) \widehat{h}_{J-1}(x) - \omega_J(x) \widehat{h}_J(x),
$$

or equivalently,

$$\sum_{j \in \mathbf{Z}} \omega_j(x) \widehat{h}_j(x) = \omega_{J-1}(x) \left[s_{J-1}(x) + s_{J-1}(2a_J - x) \right]$$
$$+ \omega_J(x) \left[s_J(x) - s_J(2a_J - x) \right].$$

But $s_J(x) = s(x) \omega_J(x)$. Therefore,

$$\sum_{j \in \mathbf{Z}} \omega_j(x) \widehat{h}_j(x) = \omega_{J-1}^2(x) s(x) + \omega_{J-1}(x) \omega_{J-1}(2a_J - x) s(2a_J - x)$$
$$+ \omega_J^2(x) s(x) - \omega_J(x) \omega_J(2a_J - x) s(2a_J - x),$$

or equivalently,

$$\sum_{j \in \mathbf{Z}} \omega_j(x) \widehat{h}_j(x) = s(x) + \omega_{J-1}(x) \omega_{J-1}(2a_J - x) s(2a_J - x)$$
$$- \omega_J(x) \omega_J(2a_J - x) s(2a_J - x).$$

Let $x = a_J + \sigma$ where $-\epsilon_J \leq \sigma \leq \epsilon_J$ then the term within the brackets becomes

$$\omega_{J-1}(a_J + \sigma) \omega_{J-1}(a_J - \sigma) - \omega_J(a_J + \sigma) \omega_J(a_J - \sigma) = 0.$$

Thus,

$$\sum_{j \in \mathbf{Z}} \omega_j(x) \widehat{h}_j(x) = s(x).$$

This proves that $\{ u_{j,k} | \ j \in \mathbf{Z}, \ k \in \mathbf{N} \}$ is a basis for $L^2(\mathbf{R})$ ∎

If the function $f_{j,k}(x)$ is chosen to be

$$f_{j,k}(x) = \sqrt{\frac{2}{a_{j+1} - a_j}} \sin \left(\pi \left(k + \frac{1}{2} \right) \left[\frac{x - a_j}{a_{j+1} - a_j} \right] \right)$$

then, the resulting basis is known as a local sinusoidal basis. An example of a window is given by

$$\omega_j(x) = \begin{cases} 0, & -\infty < x < a_j - \epsilon_j \\ \sin(\frac{\pi}{4\epsilon_j}\{x - [a_j - \epsilon_j]\}), & a_j - \epsilon_j \leq x \leq a_j + \epsilon_j \\ 1, & a_j + \epsilon_j < x < a_{j+1} - \epsilon_{j+1} \\ \cos(\frac{\pi}{4\epsilon_{j+1}}\{x - [a_{j+1} - \epsilon_{j+1}]\}), & a_{j+1} - \epsilon_{j+1} \leq x \leq a_{j+1} + \epsilon_{j+1} \\ 0, & a_{j+1} + \epsilon_{j+1} < x < \infty. \end{cases}$$

3.5 Multidimensional QMF banks

This section defines and analyzes multidimensional quadrature mirror filter (MD-QMF) banks in terms of two important formulations: alias-component

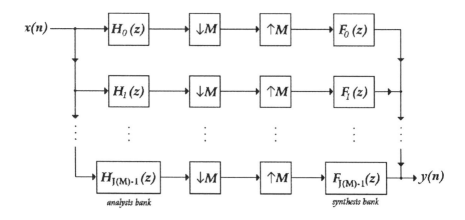

Figure 3.19. M-channel multidimensional filter bank.

and polyphase. Then, in a manner analogous to the one-dimensional case, we will closely examine the polyphase matrix, which characterizes the behavior of the multidimensional QMF banks.

For simplicity in Chapter 2, we were able to refer to shift vectors as $\mathbf{a}$ without having to assume an ordering of them. In multidimensional filter banks, we will need to associate an ordering to the shift vectors, so they will be denoted $\mathbf{a}_i$, where $\mathbf{a}_0 = [0, 0, \ldots, 0]^T$ by convention.

3.5.1 Multidimensional filter bank formulations

We will discuss two different approaches to the analysis of multidimensional filter banks: the alias-component formulation and the polyphase formulation. Then, we will show that the two formulations are equivalent.

3.5.1.1 Alias-component filter banks

The input is denoted $x(\mathbf{n})$ and the output is denoted $y(\mathbf{n})$. It consists of $J(\mathbf{M})$ multidimensional decimators; $J(\mathbf{M})$ multidimensional expanders; the $J(\mathbf{M})$ multidimensional analysis filters denoted $H_k(\mathbf{z})$, $k = 0, \ldots, J(\mathbf{M})-1$; and the $J(\mathbf{M})$ multidimensional synthesis filters denoted $F_k(\mathbf{z})$, $k = 0, \ldots, J(\mathbf{M})-1$. Figure 3.19 depicts an M-channel multidimensional filter bank.

The basic philosophy behind the design of MD-QMF banks is to permit aliasing in the multidimensional filters of the analysis bank and then choose the multidimensional filters of the synthesis bank so that the alias-

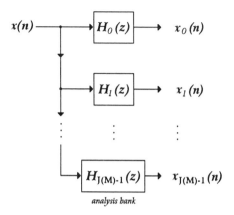

Figure 3.20. Multidimensional analysis bank.

components in the multidimensional filters of the analysis bank are cancelled.

Now, we will proceed with the analysis of the multidimensional quadrature mirror filter bank, by examining it stage-by-stage. First, we will consider the analysis bank, which is depicted by Figure 3.20. The elemental equation for this stage is given by

$$X_k(\mathbf{z}) = H_k(\mathbf{z})X(\mathbf{z}).$$

Then, we will consider the bank of decimators, which is depicted by Figure 3.21. The elemental equation for this stage is given by

$$V_k(\mathbf{z}) = \frac{1}{\mathrm{J}(\mathbf{M})} \sum_{i=0}^{\mathrm{J}(\mathbf{M})-1} X_k\left(\mathbf{z}^{\mathbf{M}^{-1}}\exp\left[-j(2\pi\mathbf{M}^{-T})\mathbf{a}_i\right]\right).$$

The following stage consists of a bank of expanders, which is depicted by Figure 3.22. The elemental equation for this stage is given by

$$U_k(\mathbf{z}) = V_k(\mathbf{z}^{\mathbf{M}}).$$

Lastly, we will consider the synthesis bank, which is depicted by Figure 3.23. The elemental equation for this stage is given by

$$Y(\mathbf{z}) = \sum_{k=0}^{\mathrm{J}(\mathbf{M})-1} F_k(\mathbf{z})U_k(\mathbf{z}).$$

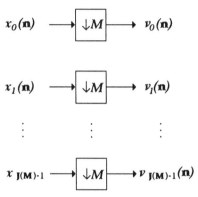

Figure 3.21. Multidimensional bank of decimators.

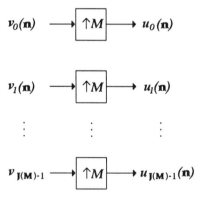

Figure 3.22. Multidimensional bank of expanders.

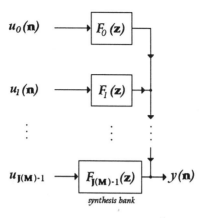

Figure 3.23. Multidimensional synthesis bank.

Combining the equations for $X_k(\mathbf{z})$, $V_k(\mathbf{z})$, and $U_k(\mathbf{z})$ yields

$$U_k(\mathbf{z}) = \frac{1}{J(\mathbf{M})} \sum_{i=0}^{J(\mathbf{M})-1} H_k(\mathbf{z} \exp[-j(2\pi\mathbf{M}^{-T})\mathbf{a}_i])X(\mathbf{z} \exp[-j(2\pi\mathbf{M}^{-T})\mathbf{a}_i]).$$

Finally, combining this equation with the equation for $Y(\mathbf{z})$ yields

$$Y(\mathbf{z}) \;=\; \underbrace{\left[\frac{1}{J(\mathbf{M})} \sum_{k=0}^{J(\mathbf{M})-1} F_k(\mathbf{z})H_k(\mathbf{z})\right] X(\mathbf{z})}_{\text{desired terms}}$$

$$+ \underbrace{\sum_{i=1}^{J(\mathbf{M})-1} A_i(\mathbf{z})\, X\left(\mathbf{z} \exp\left[-j(2\pi\mathbf{M}^{-T})\mathbf{a}_i\right]\right)}_{\text{terms due to aliasing}}$$

where,

$$A_i(\mathbf{z}) = \frac{1}{J(\mathbf{M})} \sum_{k=0}^{J(\mathbf{M})-1} F_k(\mathbf{z})H_k\left(\mathbf{z} \exp\left[-j(2\pi\mathbf{M}^{-T})\mathbf{a}_i\right]\right)$$

and $\mathbf{a}_0$ lies at the origin. A few observations can be made. First, the desired term can be interpreted as the multidimensional input signal weighted by the mean of the product of the multidimensional analysis and multidimensional synthesis filters. Secondly, if the coefficients of the aliasing term can be set to zero then a linear time invariant (LTI) system can be constructed out of linear time varying (LTV) components — multidimensional decimators and multidimensional expanders. In this case, when aliasing is cancelled the multidimensional distortion function is given by

$$T(\mathbf{z}) = \frac{1}{J(\mathbf{M})} \sum_{k=0}^{J(\mathbf{M})-1} F_k(\mathbf{z}) H_k(\mathbf{z}).$$

Unless $T(\mathbf{z})$ is allpass, *i.e.* $\mid T(\exp{(j\underline{\omega})}) \mid = c \neq 0$ for all $\underline{\omega}$, we say $Y(\mathbf{z})$ suffers from amplitude distortion. Similarly, unless $T(\mathbf{z})$ has linear phase, *i.e.* $\phi(\underline{\omega}) = \arg{(T(\exp{(j\underline{\omega})}))} = a + b\underline{\omega}$ for constant a and b, we say $Y(\mathbf{z})$ suffers from phase distortion. If the system is free from aliasing, amplitude distortion, and phase distortion, then $T(\mathbf{z})$ is a pure delay, *i.e.* $T(\mathbf{z}) = c\mathbf{z}^{-n_0}$, $c \neq 0$. In such a system, $y(\mathbf{n})$ is a scaled and delayed version of $x(\mathbf{n})$, *i.e.* $y(\mathbf{n}) = cx(\mathbf{n} - \mathbf{n}_0)$, and the resulting system is called a *perfect reconstruction* system.

Let us reexamine the output of the filter bank $Y(\mathbf{z})$, which was given by

$$Y(\mathbf{z}) = \sum_{i=0}^{J(\mathbf{M})-1} \frac{1}{J(\mathbf{M})} A_i(\mathbf{z}) \, X\left(\mathbf{z} \exp\left[-j(2\pi\mathbf{M}^{-T})\mathbf{a}_i\right]\right)$$

where

$$A_i(\mathbf{z}) = \frac{1}{J(\mathbf{M})} \sum_{k=0}^{J(\mathbf{M})-1} F_k(\mathbf{z}) H_k\left(\mathbf{z} \exp\left[-j(2\pi\mathbf{M}^{-T})\mathbf{a}_i\right]\right).$$

Writing the system of equations $A_i(\mathbf{z})$, $i = 0, \ldots, J(\mathbf{M}) - 1$, in matrix form yields

$$\mathbf{A}(\mathbf{z}) = \frac{1}{J(\mathbf{M})} \mathbf{H}(\mathbf{z}) \mathbf{f}(\mathbf{z}),$$

where the vector of gain terms $\mathbf{A}(\mathbf{z})$ is defined by

$$[\mathbf{A}(\mathbf{z})]_k = A_k(\mathbf{z}),$$

the synthesis bank $\mathbf{f}(\mathbf{z})$ is defined by

$$[\mathbf{f}(\mathbf{z})]_k = F_k(\mathbf{z}),$$

and the alias-component matrix $\mathbf{H}(\mathbf{z})$ is defined by

$$[\mathbf{H}(\mathbf{z})]_{i,k} = H_k\left(\mathbf{z} \exp\left[-j(2\pi\mathbf{M}^{-T})\mathbf{a}_i\right]\right).$$

In this formulation, aliasing can be eliminated if and only if the gain for each of the aliasing terms equals zero, that is, $A_i(\mathbf{z}) = 0$ for $i = 1, \ldots, J(\mathbf{M}) - 1$. Moreover, to insure *perfect reconstruction* $A_0(\mathbf{z})$ must be a delay, *i.e.* $A_0(\mathbf{z}) = \mathbf{z}^{-\mathbf{m}_0}$ for some integer vector $\mathbf{m}_0$.

The solution of this system of equations for $\mathbf{f}(\mathbf{z})$ may have practical difficulties. It requires the inversion of the alias-component matrix $\mathbf{H}(\mathbf{z})$. Even if successful, that is, $\mathbf{H}(\mathbf{z})$ is nonsingular, there is no guarantee that the resulting filters $\mathbf{f}(\mathbf{z})$ are stable. The approach given in the next section presents a different technique in which all of the above difficulties vanish.

3.5.1.2 Polyphase formulation of filter banks

Now, we consider the polyphase representation formulation of multidimensional filter banks. Towards this end, we will expand the *desired result* in terms of polyphase. Then, we will determine the conditions to be placed on this result so as to achieve perfect reconstruction or simply alias cancellation.

Now, recall that the desired result is given by

$$Y(\mathbf{z}) = [\frac{1}{J(\mathbf{M})} \sum_{k=0}^{J(\mathbf{M})-1} F_k(\mathbf{z}) H_k(\mathbf{z})] X(\mathbf{z}),$$

or equivalently in matrix notation

$$Y(\mathbf{z}) = \frac{1}{J(\mathbf{M})} \mathbf{f}^T(\mathbf{z}) \mathbf{h}(\mathbf{z}) X(\mathbf{z}),$$

where the synthesis bank is defined by

$$[\mathbf{f}(\mathbf{z})]_k = F_k(\mathbf{z}),$$

and the analysis bank is defined by

$$[\mathbf{h}(\mathbf{z})]_k = H_k(\mathbf{z}).$$

Writing $H_k(\mathbf{z}), k = 0, \ldots, J(\mathbf{M}) - 1$, in terms of Type-I polyphase yields

$$H_k(\mathbf{z}) = \sum_{j=0}^{J(\mathbf{M})-1} \mathbf{z}^{-\mathbf{a}_j} E_{k,\mathbf{a}_j}^{(\mathbf{M})}(\mathbf{z}^{\mathbf{M}})$$

for $\mathbf{a}_j \in \mathcal{N}(\mathbf{M})$ and $k = 0, \ldots, J(\mathbf{M}) - 1$. This system of equations can be rewritten in matrix notation as

$$\mathbf{h}(\mathbf{z}) = \mathbf{E}^{(\mathbf{M})}(\mathbf{z}^{\mathbf{M}}) \mathbf{e}_{\mathbf{M}}(\mathbf{z}),$$

where the delay chain $\mathbf{e_M(z)}$ is defined by

$$[\mathbf{e_M(z)}]_k = \mathbf{z}^{-\mathbf{a}_k}$$

and the $J(\mathbf{M})$ x $J(\mathbf{M})$ polyphase-component matrix for the multidimensional analysis bank $\mathbf{E}^{(\mathbf{M})}(\mathbf{z^M})$ is defined by

$$\left[\mathbf{E}^{(\mathbf{M})}(\mathbf{z^M})\right]_{k,j} = E^{(\mathbf{M})}_{k,\mathbf{a}_j}(\mathbf{z^M})$$

$\mathbf{a}_j \in \mathcal{N}(\mathbf{M})$ and $k = 0,\ldots, J(\mathbf{M}) - 1$. Then, writing $F_k(\mathbf{z}), k = 0,\ldots,$ $J(\mathbf{M}) - 1$ in terms of Type-II polyphase yields

$$F_k(\mathbf{z}) = \sum_{j=0}^{J(\mathbf{M})-1} \mathbf{z}^{-\mathbf{a}_{J(\mathbf{M})}-1-j} R^{(\mathbf{M})}_{\mathbf{a}_j,k}(\mathbf{z^M})$$

for $\mathbf{a}_j \in \mathcal{N}(\mathbf{M})$ and $k = 0,\ldots, J(\mathbf{M}) - 1$. This system of equations can be written in matrix notation as

$$\mathbf{f}^T(\mathbf{z}) = \mathbf{z}^{-\mathbf{a}_{J(\mathbf{M})}-1}\tilde{\mathbf{e}}_\mathbf{M}(\mathbf{z})\mathbf{R}^{(\mathbf{M})}(\mathbf{z^M}),$$

where the paraconjugation of $\mathbf{e_M(z)}$, denoted $\tilde{\mathbf{e}}_\mathbf{M}(\mathbf{z})$, is given by

$$[\tilde{\mathbf{e}}_\mathbf{M}(\mathbf{z})]_k = \mathbf{z}^{\mathbf{a}_k}$$

and the $J(\mathbf{M})$ x $J(\mathbf{M})$ polyphase-component matrix for the synthesis bank $\mathbf{R}^{(\mathbf{M})}(\mathbf{z^M})$ is defined by

$$\left[\mathbf{R}^{(\mathbf{M})}(\mathbf{z^M})\right]_{j,k} = R^{(\mathbf{M})}_{\mathbf{a}_j,k}(\mathbf{z^M})$$

$\mathbf{a}_j \in \mathcal{N}(\mathbf{M})$ and $k = 0,\ldots, J(\mathbf{M}) - 1$. Substituting these equations for $\mathbf{h(z)}$ and $\mathbf{f}^T(\mathbf{z})$ into the equation for $Y(\mathbf{z})$ yields

$$Y(\mathbf{z}) = \frac{1}{J(\mathbf{M})} \mathbf{z}^{-\mathbf{a}_{J(\mathbf{M})}-1}\tilde{\mathbf{e}}_\mathbf{M}(\mathbf{z})\mathbf{R}^{(\mathbf{M})}(\mathbf{z^M})\mathbf{E}^{(\mathbf{M})}(\mathbf{z^M})\mathbf{e_M(z)}X(\mathbf{z}).$$

Let $\mathbf{P}^{(\mathbf{M})}(\mathbf{z^M}) = \mathbf{R}^{(\mathbf{M})}(\mathbf{z^M})\mathbf{E}^{(\mathbf{M})}(\mathbf{z^M})$. Then,

$$Y(\mathbf{z}) = \frac{1}{J(\mathbf{M})} \mathbf{z}^{-\mathbf{a}_{J(\mathbf{M})}-1}\tilde{\mathbf{e}}_\mathbf{M}(\mathbf{z})\mathbf{P}^{(\mathbf{M})}(\mathbf{z^M})\mathbf{e_M(z)}X(\mathbf{z}),$$

or equivalently,

$$Y(\mathbf{z}) = T(\mathbf{z})X(\mathbf{z}),$$

where the multidimensional distortion function $T(\mathbf{z})$ is given by

$$T(\mathbf{z}) = \frac{1}{J(\mathbf{M})} \mathbf{z}^{-\mathbf{a}_{J(\mathbf{M})}-1}\tilde{\mathbf{e}}_\mathbf{M}(\mathbf{z})\mathbf{P}^{(\mathbf{M})}(\mathbf{z^M})\mathbf{e_M(z)}.$$

These equations suggest the filter bank depicted in Figure 3.24.

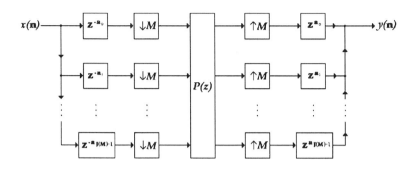

Figure 3.24. Multidimensional filter bank.

3.5.1.3 Relationship between formulations

As seen by the previous subsections, the study of filter banks can be performed using either alias-component or polyphase matrices. Now let us examine the relationship between these two formulations. The AC matrix $\mathbf{H}(\mathbf{z})$ is defined by

$$[\mathbf{H}(\mathbf{z})]_{i,m} = H_m\left(\mathbf{z} \exp\left[-j\left(2\pi\mathbf{M}^{-T}\right)\mathbf{a}_i\right]\right),$$

so,

$$\left[\mathbf{H}^T(\mathbf{z})\right]_{i,m} = [\mathbf{H}(\mathbf{z})]_{m,i}.$$

Recall that $\mathbf{h}(\mathbf{z})$ is defined by

$$[\mathbf{h}(\mathbf{z})]_m = H_m(\mathbf{z}).$$

Let the ith row of $\mathbf{H}^T(\mathbf{z})$ be designated by $\left(\mathbf{H}^T(\mathbf{z})\right)_i$. Then,

$$\left(\mathbf{H}^T(\mathbf{z})\right)_i = \mathbf{h}\left(\mathbf{z} \exp\left[-j\left(2\pi\mathbf{M}^{-T}\right)\mathbf{a}_i\right]\right).$$

Recall that $\mathbf{h}(\mathbf{z})$ is related to the polyphase-component matrix $\mathbf{E}(\mathbf{z})$ by

$$\mathbf{h}(\mathbf{z}) = \mathbf{E}^{(\mathbf{M})}(\mathbf{z}^{\mathbf{M}})\mathbf{e_M}(\mathbf{z}).$$

Therefore, $\mathbf{H}^T(\mathbf{z})$ can be defined by

$$\left(\mathbf{H}^T(\mathbf{z})\right)_i = \mathbf{E}^{(\mathbf{M})}(\mathbf{z}^{\mathbf{M}})\mathbf{e_M}(\mathbf{z} \exp\left[-j\left(2\pi\mathbf{M}^{-T}\right)\mathbf{a}_i\right]).$$

Since $[\mathbf{e_M}(\mathbf{z})]_k = \mathbf{z}^{-\mathbf{a}_k}$, then

$$\mathbf{e_M}(\mathbf{z} \exp\left[-j\left(2\pi\mathbf{M}^{-T}\right)\mathbf{a}_i\right]) = \mathbf{z}^{-\mathbf{a}_i} \exp\left[-j\left(2\pi\mathbf{M}^{-T}\right)\mathbf{a}_i\right].$$

Let $\Lambda(z) = \text{diag}[1, z^{-a_1}, \ldots, z^{-a_{J(M)-1}}]$. Then,

$$H^T(z) = E^{(M)}(z^M)\Lambda(z)W_M^H$$

where $\left[W_M^H\right]_{n,i} = \exp\left[-j\left((2\pi M^{-T})\, a_i\right)^T a_n\right]$, or equivalently,

$$H(z) = (W_M^H)^T\Lambda(z)\left(E^{(M)}(z^M)\right)^T.$$

With this equation, any results obtained in terms of the polyphase formulation can be applied to the alias-component formulation and vice versa.

3.5.2 Multidimensional alias-free filter banks

First, a background section on generalized pseudocirculant matrices is presented. This is followed by the theory of multidimensional alias-free filter banks.

3.5.2.1 Generalized pseudocirculant matrices

Definition 3.5.2.1. *Given a sampling matrix* M *and a specific ordering of shift vectors* a_i, $i = 0, \ldots, J(M) - 1$, *in* $\mathcal{N}(M)$. *Then, a generalized pseudocirculant matrix* $P(z)$ *is defined by*

$$z^{g(i,j)}P_{j,0}(z) = P_{f(i,j),i}(z)$$

where

$$g(i,j) = M^{-1}\left(a_i + a_j - a_{f(i,j)}\right)$$

and

$$f(i,j) \text{ is the integer defined by } ((a_i + a_j))_M = a_{f(i,j)}.$$

Algorithmically speaking, $P(z)$ can be determined by the following sequence of operations:

1. Evaluate $a_i + a_j$; $i=1, \ldots, J(M)-1$, $j=0, \ldots, J(M)-1$.
2. Define $f(i,j) = m$, where $((a_i + a_j))_M = a_m$ and

$$((a_i + a_j))_M = \begin{cases} a_i + a_j, & \text{if } (a_i + a_j) \in \mathcal{N}(M) \\ (a_i + a_j) - M\left\lfloor M^{-1}(a_i + a_j)\right\rfloor, & \text{otherwise.} \end{cases}$$

3. Solve for $g(i,j)$ using

$$a_i + a_j = Mg(i,j) + a_{f(i,j)}.$$

4. Evaluate the following equation

$$z^{g(i,j)}P_{j,0}(z) = P_{f(i,j),i}(z).$$

Therefore, given a sampling matrix $\mathbf{M}$ with k cosets, then all the relationships that define the generalized pseudocirculant matrix are of the form

$$\mathbf{z}^{\mathbf{g}(i,j)} P_{j,0}(\mathbf{z}) = P_{f(i,j),i}(\mathbf{z}) \; ; \text{ where } i,j,f(i,j) \in \{0,1,\ldots,k-1\}.$$

Thus, $\mathbf{P}(\mathbf{z})$ is a $k \times k$ matrix, since $\dim(\mathcal{N}(\mathbf{M})) = k$.

For example, let us determine the generalized pseudocirculant matrix with sampling matrix

$$\mathbf{M} = \begin{bmatrix} 1 & 1 \\ -1 & 2 \end{bmatrix}$$

with cosets given by

$$\mathbf{a}_0 = \begin{bmatrix} 0 \\ 0 \end{bmatrix} \; ; \; \mathbf{a}_1 = \begin{bmatrix} 1 \\ 0 \end{bmatrix} \; ; \text{ and } \mathbf{a}_2 = \begin{bmatrix} 1 \\ 1 \end{bmatrix}.$$

In addition, note that $\mathbf{M}^{-1} = \frac{1}{3} \begin{bmatrix} 2 & -1 \\ 1 & 1 \end{bmatrix}$.

Case I $(\mathbf{a}_0 + \mathbf{a}_1)$:

(1) Evaluate $\mathbf{a}_0 + \mathbf{a}_1 = \begin{bmatrix} 0 \\ 0 \end{bmatrix} + \begin{bmatrix} 1 \\ 0 \end{bmatrix} = \begin{bmatrix} 1 \\ 0 \end{bmatrix} = \mathbf{a}_1.$

(2) Since $\mathbf{a}_0 + \mathbf{a}_1 \in \mathcal{N}(\mathbf{M})$, then $((\mathbf{a}_0 + \mathbf{a}_1))_{\mathbf{M}} = \mathbf{a}_0 + \mathbf{a}_1 = \mathbf{a}_1.$

Thus, $f(1,0) = 1.$

(3) Since $\mathbf{a}_0 + \mathbf{a}_1 = \mathbf{a}_{f(1,0)} = \mathbf{a}_1$, then $\mathbf{Mg}(1,0) = \begin{bmatrix} 0 \\ 0 \end{bmatrix}.$

Hence, $\mathbf{g}(1,0) = \begin{bmatrix} 0 \\ 0 \end{bmatrix}.$

(4) Therefore,

$$z_0^0 z_1^0 P_{0,0}(\mathbf{z}) = P_{1,1}(\mathbf{z}),$$

or equivalently,

$$P_{0,0}(\mathbf{z}) = P_{1,1}(\mathbf{z}).$$

Case II $(\mathbf{a}_0 + \mathbf{a}_2)$:

(1) Evaluate $\mathbf{a}_0 + \mathbf{a}_2 = \begin{bmatrix} 0 \\ 0 \end{bmatrix} + \begin{bmatrix} 1 \\ 1 \end{bmatrix} = \begin{bmatrix} 1 \\ 1 \end{bmatrix} = \mathbf{a}_2.$

(2) Since $\mathbf{a}_0 + \mathbf{a}_2 \in \mathcal{N}(\mathbf{M})$, then $((\mathbf{a}_0 + \mathbf{a}_2))_{\mathbf{M}} = \mathbf{a}_0 + \mathbf{a}_2 = \mathbf{a}_2.$

Thus, $f(2,0) = 2$.

(3) Since $\mathbf{a}_0 + \mathbf{a}_2 = \mathbf{a}_{f(2,0)} = \mathbf{a}_2$, then $\mathbf{Mg}(2,0) = \begin{bmatrix} 0 \\ 0 \end{bmatrix}$.

Hence, $\mathbf{g}(2,0) = \begin{bmatrix} 0 \\ 0 \end{bmatrix}$.

(4) Therefore,

$$z_0^0 z_1^0 P_{0,0}(\mathbf{z}) = P_{2,2}(\mathbf{z}),$$

or equivalently,

$$P_{0,0}(\mathbf{z}) = P_{2,2}(\mathbf{z}).$$

Case III $(\mathbf{a}_1 + \mathbf{a}_1)$:

(1) Evaluate $\mathbf{a}_1 + \mathbf{a}_1 = \begin{bmatrix} 1 \\ 0 \end{bmatrix} + \begin{bmatrix} 1 \\ 0 \end{bmatrix} = \begin{bmatrix} 2 \\ 0 \end{bmatrix}$.

(2) Since $\mathbf{a}_1 + \mathbf{a}_1 \notin \mathcal{N}(\mathbf{M})$, then

$$((\mathbf{a}_1 + \mathbf{a}_1))_{\mathbf{M}} = \begin{bmatrix} 2 \\ 0 \end{bmatrix} - \begin{bmatrix} 1 & 1 \\ -1 & 2 \end{bmatrix} \left| \frac{1}{3} \begin{bmatrix} 2 & -1 \\ 1 & 1 \end{bmatrix} \begin{bmatrix} 2 \\ 0 \end{bmatrix} \right| = \mathbf{a}_2.$$

Thus, $f(1,1) = 2$.

(3) Solve for $\mathbf{g}(1,1)$:

$$\mathbf{a}_1 + \mathbf{a}_1 = \mathbf{Mg}(1,1) + \mathbf{a}_{f(1,1)} \Rightarrow \mathbf{Mg}(1,1) = \begin{bmatrix} 1 \\ -1 \end{bmatrix}.$$

Hence, $\mathbf{g}(1,1) = \frac{1}{3} \begin{bmatrix} 2 & -1 \\ 1 & 1 \end{bmatrix} \begin{bmatrix} 1 \\ -1 \end{bmatrix} = \begin{bmatrix} 1 \\ 0 \end{bmatrix}$.

(4) Therefore,

$$z_0^1 z_1^0 P_{1,0}(\mathbf{z}) = P_{2,1}(\mathbf{z}),$$

or equivalently,

$$z_0 P_{1,0}(\mathbf{z}) = P_{2,1}(\mathbf{z}).$$

Case IV $(\mathbf{a}_1 + \mathbf{a}_2)$:

(1) Evaluate $\mathbf{a}_1 + \mathbf{a}_2 = \begin{bmatrix} 1 \\ 0 \end{bmatrix} + \begin{bmatrix} 1 \\ 1 \end{bmatrix} = \begin{bmatrix} 2 \\ 1 \end{bmatrix}$.

(2) Since $\mathbf{a}_1 + \mathbf{a}_2 \notin \mathcal{N}(\mathbf{M})$, then

$$((\mathbf{a}_1 + \mathbf{a}_2))_{\mathbf{M}} = \begin{bmatrix} 2 \\ 1 \end{bmatrix} - \begin{bmatrix} 1 & 1 \\ -1 & 2 \end{bmatrix} \left\lfloor \frac{1}{3} \begin{bmatrix} 2 & -1 \\ 1 & 1 \end{bmatrix} \begin{bmatrix} 2 \\ 1 \end{bmatrix} \right\rfloor = \mathbf{a}_0.$$

Thus, $f(1,2) = f(2,1) = 0$.

(3) Solve for $\mathbf{g}(1,2)$:

$$\mathbf{a}_1 + \mathbf{a}_2 = \mathbf{Mg}(1,2) + \mathbf{a}_{f(1,2)} \Rightarrow \mathbf{Mg}(1,2) = \begin{bmatrix} 2 \\ 1 \end{bmatrix}.$$

Hence, $\mathbf{g}(1,2) = \frac{1}{3} \begin{bmatrix} 2 & -1 \\ 1 & 1 \end{bmatrix} \begin{bmatrix} 2 \\ 1 \end{bmatrix} = \begin{bmatrix} 1 \\ 1 \end{bmatrix}.$

Similarly, $\mathbf{g}(2,1) = \begin{bmatrix} 1 \\ 1 \end{bmatrix}.$

(4) Therefore,

$$z_0^1 z_1^1 P_{2,0}(\mathbf{z}) = P_{0,1}(\mathbf{z}),$$

or equivalently,

$$z_0 z_1 P_{2,0}(\mathbf{z}) = P_{0,1}(\mathbf{z}).$$

In addition,

$$z_0^1 z_1^1 P_{1,0}(\mathbf{z}) = P_{0,2}(\mathbf{z}),$$

or equivalently,

$$z_0 z_1 P_{1,0}(\mathbf{z}) = P_{0,2}(\mathbf{z}).$$

Case V $(\mathbf{a}_2 + \mathbf{a}_2)$:

(1) Evaluate $\mathbf{a}_2 + \mathbf{a}_2 = \begin{bmatrix} 1 \\ 1 \end{bmatrix} + \begin{bmatrix} 1 \\ 1 \end{bmatrix} = \begin{bmatrix} 2 \\ 2 \end{bmatrix}.$

(2) Since $\mathbf{a}_2 + \mathbf{a}_2 \notin \mathcal{N}(\mathbf{M})$, then

$$((\mathbf{a}_2 + \mathbf{a}_2))_{\mathbf{M}} = \begin{bmatrix} 2 \\ 2 \end{bmatrix} - \begin{bmatrix} 1 & 1 \\ -1 & 2 \end{bmatrix} \left\lfloor \frac{1}{3} \begin{bmatrix} 2 & -1 \\ 1 & 1 \end{bmatrix} \begin{bmatrix} 2 \\ 2 \end{bmatrix} \right\rfloor = \mathbf{a}_1.$$

Thus, $f(2,2) = 1$.

(3) Solve for $\mathbf{g}(2,2)$:

$$\mathbf{a}_2 + \mathbf{a}_2 = \mathbf{Mg}(2,2) + \mathbf{a}_{f(2,2)} \Rightarrow \mathbf{Mg}(2,2) = \begin{bmatrix} 1 \\ 2 \end{bmatrix}.$$

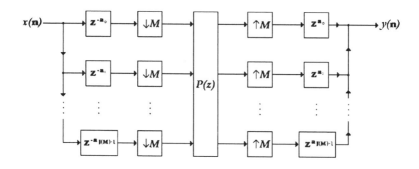

Figure 3.25. Multidimensional filter bank.

Hence, $\mathbf{g}(2,2) = \frac{1}{3} \begin{bmatrix} 2 & -1 \\ 1 & 1 \end{bmatrix} \begin{bmatrix} 1 \\ 2 \end{bmatrix} = \begin{bmatrix} 0 \\ 1 \end{bmatrix}.$

(4) Therefore,

$$z_0^0 z_1^1 P_{2,0}(\mathbf{z}) = P_{1,2}(\mathbf{z}),$$

or equivalently,

$$z_1 P_{2,0}(\mathbf{z}) = P_{1,2}(\mathbf{z}).$$

With the conditions that were obtained by considering the above cases, the following generalized pseudocirculant matrix is obtained.

$$\mathbf{P}(\mathbf{z}) = \begin{bmatrix} P_0(\mathbf{z}) & z_0 z_1 P_2(\mathbf{z}) & z_0 z_1 P_1(\mathbf{z}) \\ P_1(\mathbf{z}) & P_0(\mathbf{z}) & z_1 P_2(\mathbf{z}) \\ P_2(\mathbf{z}) & z_0 P_1(\mathbf{z}) & P_0(\mathbf{z}) \end{bmatrix}.$$

3.5.2.2 Alias-free filter banks

Consider the following multidimensional filter bank depicted in Figure 3.25. In the polyphase representation of a one-dimensional filter bank, we utilize delay elements at both ends of the filter bank, since there is no need to make a noncausal filter bank because of delays. On the other hand, in a multidimensional filter bank, we will shift the inputs and shift them back at the outputs. We observe that $Y(\mathbf{z}) =$

$$\frac{1}{\mathbf{J(M)}} \sum_{l=0}^{\mathbf{J(M)}-1} X(\mathbf{z} \exp[-j2\pi \mathbf{M}^{-T} \mathbf{a}_l]) \sum_{m=0}^{\mathbf{J(M)}-1} \exp[-j2\pi \mathbf{M}^{-T} \mathbf{a}_l]^{-\mathbf{a}_m} V_m(\mathbf{z})$$

where

$$V_m(\mathbf{z}) = \sum_{s=0}^{\mathbf{J(M)}-1} \mathbf{z}^{-\mathbf{a}_m} \mathbf{z}^{\mathbf{a}_s} P_{s,m}(\mathbf{z}^{\mathbf{M}}).$$

Since $X(\mathbf{z} \exp[-j(2\pi\mathbf{M}^{-T})\mathbf{k}_l])$, $l \neq 0$, represents the alias terms, then the resulting equation for $Y(\mathbf{z})$ is free from aliasing if and only if

$$\sum_{m=0}^{\mathbf{J(M)}-1} \exp[-j(2\pi\mathbf{M}^{-T})\mathbf{a}_l]^{-\mathbf{a}_m} V_m(\mathbf{z}) = 0 \text{ for } l \neq 0,$$

or equivalently

$$\sum_{m=0}^{\mathbf{J(M)}-1} \exp[j2\pi\mathbf{a}_l^T\mathbf{M}^{-1}\mathbf{a}_m] V_m(\mathbf{z}) = 0 \text{ for } l \neq 0.$$

But $\exp[j2\pi\mathbf{k}_l^T\mathbf{M}^{-1}\mathbf{k}_m]$ is simply the (l,m)th element of the complex conjugate transposed of the generalized discrete Fourier transform matrix, $\mathbf{W}_{\mathbf{M}}^{(g)}$. Hence,

$$\left[\mathbf{W}_{\mathbf{M}}^{(g)}\right]^H
\begin{bmatrix} V_0(\mathbf{z}) \\ V_1(\mathbf{z}) \\ \vdots \\ V_{\mathbf{J(M)}-1}(\mathbf{z}) \end{bmatrix}
=
\begin{bmatrix} * \\ 0 \\ \vdots \\ 0 \end{bmatrix}.$$

Premultiply both sides of this equation by $\mathbf{W}_{\mathbf{M}}^{(g)}$. Then, using the fact that $\mathbf{W}_{\mathbf{M}}^{(g)} \left[\mathbf{W}_{\mathbf{M}}^{(g)}\right]^H = \mathbf{J(M)I}$, this equation becomes

$$\mathbf{J(M)}
\begin{bmatrix} V_0(\mathbf{z}) \\ V_1(\mathbf{z}) \\ \vdots \\ V_{\mathbf{J(M)}-1}(\mathbf{z}) \end{bmatrix}
= \mathbf{W}_{\mathbf{M}}^{(g)}
\begin{bmatrix} * \\ 0 \\ \vdots \\ 0 \end{bmatrix}.$$

Since all the entries in the first column of $\mathbf{W}_{\mathbf{M}}^{(g)}$ are equal to one, then

$$V_0(\mathbf{z}) = V_1(\mathbf{z}) = \cdots = V_{\mathbf{J(M)}-1}(\mathbf{z}).$$

Therefore, let

$$V(\mathbf{z}) = V_m(\mathbf{z}) \text{ for } m = 0, \ldots, \mathbf{J(M)} - 1.$$

Hence,

$$z^{\mathbf{a}_i} \sum_{j=0}^{J(\mathbf{M})-1} z^{\mathbf{a}_j} P_{j,0}(z^{\mathbf{M}}) = \sum_{l=0}^{J(\mathbf{M})-1} z^{\mathbf{a}_l} P_{l,i}(z^{\mathbf{M}})$$

for all $i = 0, \ldots, J(\mathbf{M}) - 1$, or equivalently,

$$\sum_{j=0}^{J(\mathbf{M})-1} z^{\mathbf{a}_j + \mathbf{a}_i} P_{j,0}(z^{\mathbf{M}}) = \sum_{l=0}^{J(\mathbf{M})-1} z^{\mathbf{a}_l} P_{l,i}(z^{\mathbf{M}}),$$

for all $i = 0, \ldots, J(\mathbf{M}) - 1$. Using the Multidimensional Division Theorem, we can express $\mathbf{a}_j + \mathbf{a}_i$ as

$$\mathbf{a}_j + \mathbf{a}_i = \mathbf{Mg}(i,j) + \mathbf{a}_{f(i,j)}$$

where $\mathbf{g}(i,j) \in \mathcal{N}$ and $\mathbf{a}_{f(i,j)} \in \mathcal{N}(\mathbf{M})$. Then,

$$\sum_{j=0}^{J(\mathbf{M})-1} z^{\mathbf{Mg}(i,j)} z^{\mathbf{a}_{f(i,j)}} P_{j,0}(z^{\mathbf{M}}) = \sum_{l=0}^{J(\mathbf{M})-1} z^{\mathbf{a}_l} P_{l,i}(z^{\mathbf{M}})$$

for all $i = 0, \ldots, J(\mathbf{M}) - 1$. This leads to

$$z^{\mathbf{g}(i,j)} P_{j,0}(z) = P_{f(i,j),i}(z)$$

for all $i, j = 0, \ldots, J(\mathbf{M}) - 1$. Polynomial matrices satisfying this relation are called generalized pseudocirculant matrices with respect to $\mathbf{M}$ for a specific ordering of $\mathbf{a}_i$s in $\mathcal{N}(\mathbf{M})$.

3.5.2.3 Perfect reconstruction QMF bank

The multidimensional filter bank achieves perfect reconstruction if and only if $\mathbf{P}(z)$ is a generalized pseudocirculant matrix and all the elements in the first column are zero except the one $P_{j,0}(z)$ that is equal to a delay, that is, if $j = j_0$ is the index of the nonzero entry, then

$$P_{j,0}(z) = \begin{cases} cz^{\mathbf{M}} & \text{if } j = j_0 \\ 0 & \text{otherwise.} \end{cases}$$

3.6 Problems

1. Consider a two-channel alias-free filter bank. Using the alias-component formulation, solve for the most general form of synthesis filters. If we assume perfect reconstruction, then how do these equations simplify?

2. Let $x(n)$ be an arbitrary sequence and let $x_1(n)$ be the first divided difference, that is,

$$x_1(n) = x(n) - x(n-1),$$

and let $x_2(n)$ be the sum of the last two samples, that is,

$$x_2(n) = x(n) + x(n-1).$$

Consider the two sequences $y_1(n)$ and $y_2(n)$ that are defined by

$$y_1(n) = x_1(2n) \quad \text{and} \quad y_2(n) = x_2(2n).$$

Can we recover $x(n)$ from $y_1(n)$ and $y_2(n)$?

3. Consider a special case of the Lloyd-Max quantizer, where the probability density function is a constant over the L-transition levels of the quantizer. What is the probability density function? What is the corresponding mean square error?

4. Let $A(z)$ and $B(z)$ be $k \times k$ pseudocirculant matrices. Prove that $A(z)$ commutes with $B(z)$, that is,

$$A(z)B(z) = B(z)A(z).$$

5. Construct the generalized pseudocirculant matrix using the sampling matrix

$$\mathbf{M} = \begin{bmatrix} 1 & 1 \\ 1 & -1 \end{bmatrix}$$

and with cosets defined by

$$\mathbf{a}_0 = \begin{bmatrix} 0 \\ 0 \end{bmatrix} \text{ and } \mathbf{a}_1 = \begin{bmatrix} 1 \\ 0 \end{bmatrix}.$$

Chapter 4

Lattice Structures

4.1 Introduction

This chapter introduces the notion of lattice structures for the realization of filter banks. The origin of lattice structures for continuous lossless systems was a classical theory of LC circuit networks, since they do not generate or dissipate energy (see, for example, Belevitch[2]). Results on discrete time systems and their factorization can be found in Vaidyanathan *et al.*[51] and Doğanata and Vaidyanathan[14]. Some of the concepts developed in this chapter are also discussed in the texts by Fliege[19], by Strang and Nguyen[46], and by Vaidyanathan[49].

Section 4.2 introduces multi-input multi-output (MIMO) linear system theory. Section 4.3 presents lattice structures for lossless systems.

4.2 Framework for lattice structures

This section systematically presents concepts that act as a framework for our study of lattice structures. These concepts include an introduction to multi-input multi-output systems, the Smith-McMillan form, and the McMillan degree of a system.

4.2.1 Multi-input multi-output systems

Definition 4.2.1.1. Let $\mathbf{u}(n)$ be an input vector of length r, that is,

$$\mathbf{u}(n) = [u_0(n), \ldots, u_{r-1}(n)]^T$$

and let $\mathbf{y}(n)$ be an output vector of length p, that is,

$$\mathbf{y}(n) = [y_0(n), \ldots, y_{p-1}(n)]^T.$$

145

Then, the multi-input multi-output system is described by

$$\mathbf{Y}(z) = \mathbf{H}(z)\mathbf{U}(z)$$

where,

$$\mathbf{U}(z) = \sum_{n=-\infty}^{\infty} \mathbf{u}(n)\, z^{-n}$$

$$\mathbf{Y}(z) = \sum_{n=-\infty}^{\infty} \mathbf{y}(n)\, z^{-n}$$

and $[\mathbf{H}(z)]_{km}$ *denotes the transfer function from the* mth *input to the* kth *output.*

The matrix $\mathbf{H}(z)$ is called the transfer matrix of the system. We will use the term r-input p-output system and $p \times r$ system, interchangeably.

4.2.1.1   Lossless systems

An important class of MIMO systems are lossless systems.

Definition 4.2.1.2. *Let* $\mathbf{H}(z)$ *be a* $p \times r$ *system.* $\mathbf{H}(z)$ *is said to be lossless if (a) each entry* $[\mathbf{H}(z)]_{km}$ *is stable and (b)* $\mathbf{H}(z)$ *is unitary on the unit circle, that is,*

$$\mathbf{H}^H(\exp(j\omega))\mathbf{H}(\exp(j\omega)) = c\mathbf{I}_r$$

for all $\omega \in [0, 2\pi)$ *and some real constant* c. *If* $c = 1$, *then* $\mathbf{H}(z)$ *is known as a normalized lossless system.*

Since all FIR systems are stable, references to FIR systems tend to use the words *paraunitary* and *lossless* interchangeably.

4.2.1.2   Impulse response matrix

Let $h_{km}(n)$ denote the impulse response of the transfer function $H_{km}(z)$. In addition, let $\mathbf{h}(n)$ represent the $p \times r$ matrix of impulse responses, where $h_{km}(n) = [\mathbf{h}(n)]_{km}$. Then,

$$\mathbf{H}(z) = \sum_{n=-\infty}^{\infty} \mathbf{h}(n)\, z^{-n}$$

where $\mathbf{H}(z)$ and $\mathbf{h}(n)$ are $p \times r$ matrices. The matrix sequence $\mathbf{h}(n)$ is called the impulse response of the system. To illustrate this idea, let

$$\mathbf{H}(z) = \begin{bmatrix} 1 + z^{-1} + z^{-2} & 2 + z^{-1} \\ 1 - z^{-1} & 2 \end{bmatrix}.$$

Then, this can be written as

$$\mathbf{H}(z) = \underbrace{\begin{bmatrix} 1 & 2 \\ 1 & 2 \end{bmatrix}}_{\mathbf{h}(0)} + \underbrace{\begin{bmatrix} 1 & 1 \\ -1 & 0 \end{bmatrix}}_{\mathbf{h}(1)} z^{-1} + \underbrace{\begin{bmatrix} 1 & 0 \\ 0 & 0 \end{bmatrix}}_{\mathbf{h}(2)} z^{-2}.$$

The following theorem provides an important property of the impulse response matrices of lossless systems.

Theorem 4.2.1.1. *Given* $\mathbf{P}(z) = \sum_{n=0}^{N} \mathbf{p}(n) z^{-n}$. *If* $\mathbf{P}(z)$ *is paraunitary, then* $\mathbf{p}^H(0)\mathbf{p}(N) = \mathbf{p}^H(N)\mathbf{p}(0) = \mathbf{0}$.

Proof: Apply the definition of paraunitariness to $\mathbf{P}(z)$ to yield

$$\widetilde{\mathbf{P}}(z)\mathbf{P}(z) = \left(\sum_{k=0}^{N} \mathbf{p}^H(k)z^k \right) \left(\sum_{n=0}^{N} \mathbf{p}(n)z^{-n} \right)$$

or equivalently,

$$\widetilde{\mathbf{P}}(z)\mathbf{P}(z) = \sum_{k=0}^{N} \sum_{n=0}^{N} \mathbf{p}^H(k)\mathbf{p}(n)z^{-(n-k)}.$$

Since $\mathbf{P}(z)$ is paraunitary, then $\widetilde{\mathbf{P}}(z)\mathbf{P}(z) = c\mathbf{I}$. Therefore,

$$\mathbf{p}^H(N)\mathbf{p}(0)z^N + \cdots + \left[\sum_{j=0}^{N} \mathbf{p}^H(j)\mathbf{p}(j) \right] z^0 + \cdots + \mathbf{p}^H(0)\mathbf{p}(N)z^{-N} = c\mathbf{I}.$$

Since $\widetilde{\mathbf{P}}(z)\mathbf{P}(z)$ will be nonzero only for coefficients of z^0, then $\mathbf{p}^H(N)\mathbf{p}(0) = \mathbf{p}^H(N)\mathbf{p}(0) = \mathbf{0}$. ∎

4.2.1.3 Polynomial matrices

Matrix polynomials play an important role in MIMO systems.

Definition 4.2.1.3. *A* $p \times r$ *polynomial matrix* $\mathbf{P}(z)$ *in variable* z *is a* $p \times r$ *matrix whose entries are polynomials in* z. *The matrix can be expressed as*

$$\mathbf{P}(z) = \sum_{n=0}^{k} \mathbf{p}(n) z^n.$$

If $\mathbf{p}(k)$ *is not the zero matrix, then* k *is called the* order *of the polynomial matrix. For example, a causal FIR system is a polynomial matrix in* z^{-1} *with order* k, *that is,*

$$\mathbf{H}(z) = \sum_{n=0}^{k} \mathbf{h}(n) z^{-n}.$$

4.2.1.4 Unimodular polynomial matrices

Definition 4.2.1.4. *A unimodular polynomial matrix* $\mathbf{U}(z)$ *in variable* z *is a square polynomial matrix in* z *with a constant nonzero determinant.*

To illustrate unimodular polynomial matrices, consider the following examples:

$$\mathbf{U}_1(z) = \begin{bmatrix} 1 & 0 \\ 1+z^3 & 1 \end{bmatrix} \text{ is unimodular, because det } (\mathbf{U}_1(z)) = 1.$$

and

$$\mathbf{U}_2(z) = \begin{bmatrix} 1+z^2 & z \\ 2z & 2 \end{bmatrix} \text{ is unimodular, because det } (\mathbf{U}_2(z)) = 2.$$

The following theorems provide properties of these matrices.

Theorem 4.2.1.2. *If* $\mathbf{A}$ *is a unimodular polynomial matrix, then* $\mathbf{A}^{-1}$ *exists and is a unimodular polynomial matrix.*

Proof: Let $\mathbf{A}$ be a unimodular polynomial matrix. Since $\det \mathbf{A} \neq 0$, then $\mathbf{A}^{-1}$ exists and $\mathbf{A}\mathbf{A}^{-1} = \mathbf{I}$. Since $\det \mathbf{A} \det \mathbf{A}^{-1} = 1$ and $|\det \mathbf{A}| = c$, then $|\det \mathbf{A}^{-1}| = \frac{1}{c}$. Since $\det \mathbf{A}\mathbf{B} = (\det \mathbf{A})(\det \mathbf{B})$, then $(\det \mathbf{A})(\det \mathbf{A}^{-1})$ $= 1$. Since $\mathbf{A}$ is unimodular, then $|\det \mathbf{A}| = c$. Hence, $|\det \mathbf{A}^{-1}| = \frac{1}{c}$. Therefore, $\mathbf{A}^{-1}$ is a unimodular polynomial matrix. ∎

Theorem 4.2.1.3. *If* $\mathbf{A}$ *and* $\mathbf{B}$ *are unimodular polynomial matrices, then* $\mathbf{A}\mathbf{B}$ *is a unimodular polynomial matrix.*

Proof: Since $\mathbf{A}$ and $\mathbf{B}$ are unimodular matrices, then $|\det \mathbf{A}| = c$ and $|\det \mathbf{B}| = d$. Since $\det \mathbf{A}\mathbf{B} = (\det \mathbf{A})(\det \mathbf{B})$, then $|\det \mathbf{A}\mathbf{B}| = cd$. Therefore, $\mathbf{A}\mathbf{B}$ is a unimodular polynomial matrix. ∎

4.2.1.5 Rank of a polynomial matrix

Definition 4.2.1.5. *The rank of a polynomial matrix is defined as the dimension of the submatrix that corresponds to the largest nonzero determinantal minor.*

Clearly, if $\mathbf{P}(z)$ is a p x r polynomial matrix, then

$$\text{rank}\,(\mathbf{P}(z)) \leq \ min\{p, r\}.$$

4.2.2   Smith-McMillan form

The theory of the Smith-McMillan form is developed for causal Linear Time Invariant(LTI) systems. The theory of the Smith form for polynomial matrices is presented first in order to provide the necessary background for the Smith-McMillan form.

The analysis of the Smith form, that is developed in this chapter for polynomial matrices, is analogous to the theory developed for integer matrices in Section 2.2. This is true because the set of polynomial matrices and the set of integer matrices belong to a common algebraic structure called the *principal ideal domain*.

4.2.2.1   Elementary operations

Elementary row (or column) operations on polynomial matrices are important because they permit the patterning of polynomial matrices into simpler forms, such as triangular and diagonal forms.

Definition 4.2.2.1. *An elementary row operation on a polynomial matrix* $\mathbf{P}(z)$ *is defined to be any of the following:*

Type-1:   Interchange two rows.

Type-2:   Multiply a row by a nonzero constant c.

Type-3:   Add a polynomial multiple of a row to another row.

These operations can be represented by premultiplying $\mathbf{P}(z)$ with an appropriate square matrix , called an elementary matrix. To illustrate these elementary operations, consider the following examples. (By convention, the rows and columns are numbered starting with zero rather than one.) The first example is a Type-1 elementary matrix that interchanges row 0 and row 3, which has the form

$$\begin{bmatrix} 0 & 0 & 0 & 1 \\ 0 & 1 & 0 & 0 \\ 0 & 0 & 1 & 0 \\ 1 & 0 & 0 & 0 \end{bmatrix}.$$

The second example is a Type-2 elementary matrix that multiplies elements in row 1 by $c \neq 0$, which has the form

$$\begin{bmatrix} 1 & 0 & 0 & 0 \\ 0 & c & 0 & 0 \\ 0 & 0 & 1 & 0 \\ 0 & 0 & 0 & 1 \end{bmatrix}.$$

The third example is a Type-3 elementary matrix that replaces row 3 with row 3 + ($a(z)$ * row0), which has the form

$$\begin{bmatrix} 1 & 0 & 0 & 0 \\ 0 & 1 & 0 & 0 \\ 0 & 0 & 1 & 0 \\ a(z) & 0 & 0 & 1 \end{bmatrix}.$$

All three types of elementary polynomial matrices are unimodular polynomial matrices. Elementary column operations are defined in a similar way by postmultiplying $\mathbf{P}(z)$ with the appropriate square matrix.

These elementary operations can be used to diagonalize a polynomial matrix. The key theorem which enables us to obtain this diagonalization is the Division Theorem for Polynomials. It states that if $N(z)$ and $D(z)$ are polynomials in z, where the order of $N(z) \geq$ order of $D(z)$, then there exists unique polynomials $Q(z)$ and $R(z)$ such that

$$N(z) = Q(z)D(z) + R(z)$$

where the order of $R(z) <$ order of $D(z)$.

4.2.2.2 Smith form decomposition

Theorem 4.2.2.1. *Every polynomial matrix* $\mathbf{A}(z)$ *can be expressed in its corresponding Smith form decomposition as*

$$\mathbf{A}(z) = \mathbf{U}(z)\mathbf{S}(z)\mathbf{V}(z)$$

where $\mathbf{U}(z), \mathbf{V}(z)$ *are unimodular polynomial matrices and the Smith form* $\mathbf{S}(z)$ *is given by*

$$\mathbf{S}(z) = \text{diag}\left(s_0(z), \ldots, s_{r-1}(z), 0, 0, \ldots, 0\right)$$

where r *is the rank of* $\mathbf{A}(z)$ *and* $s_i(z)|s_{i+1}(z)$, $i = 0, \ldots, r - 2$.

Proof: Assume that the zeroth column of $\mathbf{A}(z)$ contains a nonzero element, which may be brought to the (0,0) position by elementary operations. This element is the gcd of the zeroth column. If the new (0,0) element does not divide all the elements in the zeroth row, then it may be replaced by the gcd of the elements of the zeroth row (the effect will be that it will contain fewer prime factors than before). The process is repeated until an element in the (0,0) position is obtained which divides every element of the zeroth row and column. By elementary row and column operations, all the elements in the zeroth row and column, other than the (0,0) element, may be made zero. Denote this new submatrix formed by deleting the zeroth row and zeroth column by $\mathbf{C}(z)$.

Suppose that the submatrix of $\mathbf{C}(z)$ contains an element $c_{i,j}(z)$ which is not divisible by $c_{00}(z)$. Add column j to column 0. Column 0 then consists of elements $c_{00}, c_{1j}, \ldots, c_{n-1,j}$. Repeating the previous process, we replace c_{00} by a proper divisor of itself using elementary operations. Then, we must finally reach the stage where the element in the (0,0) position divides every element of the matrix, and all other elements of the zeroth row and column are zero.

The entire process is repeated with the submatrix obtained by deleting the zeroth row and column. Eventually a stage is reached when the matrix has the form

$$
\left[\begin{array}{cc} \mathbf{D}(z) & \mathbf{0} \\ \mathbf{0} & \mathbf{E}(z) \end{array} \right]
$$

where $\mathbf{D}(z) = \text{diag}\,(s_0(z), \ldots, s_{r-1}(z))$ and $s_i(z)|s_{i+1}(z)$, $i = 0, \ldots, r - 2$. But $\mathbf{E}(z)$ must be the zero matrix, since otherwise $\mathbf{A}(z)$ would have a rank larger than r. ∎

By convention the polynomials $s_i(z)$, $i = 0, \ldots, r - 1$, are monic polynomials, that is the highest power of the polynomial has a coefficient of unity. Note that although the two unimodular polynomial matrices $\mathbf{U}(z)$ and $\mathbf{V}(z)$ are, in general, not unique, the diagonal matrix $\mathbf{S}(z)$ is uniquely determined by $\mathbf{A}(z)$.

Example 4.2.2.1. To illustrate the Smith form decomposition, consider the following example. Let

$$
\mathbf{A}(z) = \left[\begin{array}{cc} z + 1 & z \\ 2z^2 + 3 & 2(z + 1)^2 \end{array} \right].
$$

If we divide the (1,0) element, $2z^2 + 3$, by the (0,0) element, $z + 1$, we obtain

$$
2z^2 + 3 = \underbrace{2(z - 1)}_{\text{quotient}} (z + 1) + \underbrace{5}_{\text{remainder}}.
$$

Therefore, if we apply a Type-3 row operation, which is defined by

$$
\left[\begin{array}{cc} 1 & 0 \\ -2(z - 1) & 1 \end{array} \right]
$$

to $\mathbf{A}(z)$, we will reduce the (1,0) element to the constant value of 5. Therefore,

$$
\left[\begin{array}{cc} 1 & 0 \\ -2(z - 1) & 1 \end{array} \right] \left[\begin{array}{cc} z + 1 & z \\ 2z^2 + 3 & 2(z + 1)^2 \end{array} \right] = \left[\begin{array}{cc} z + 1 & z \\ 5 & 2(3z + 1) \end{array} \right].
$$

Reduce the $(0,0)$ element to a constant with a Type-3 row operation, which is defined by $\begin{bmatrix} 1 & -\frac{z}{5} \\ 0 & 1 \end{bmatrix}$. Then, we obtain

$$\begin{bmatrix} 1 & -\frac{z}{5} \\ 0 & 1 \end{bmatrix} \begin{bmatrix} z+1 & z \\ 5 & 2(3z+1) \end{bmatrix} = \begin{bmatrix} 1 & \frac{3z(1-2z)}{5} \\ 5 & 2(3z+1) \end{bmatrix}.$$

Transform the $(0,1)$ element to zero by a Type-3 column operation, which is defined by $\begin{bmatrix} 1 & -\frac{3z(1-2z)}{5} \\ 0 & 1 \end{bmatrix}$. Then, we obtain

$$\begin{bmatrix} 1 & \frac{3z(1-2z)}{5} \\ 5 & 2(3z+1) \end{bmatrix} \begin{bmatrix} 1 & -\frac{3z(1-2z)}{5} \\ 0 & 1 \end{bmatrix} = \begin{bmatrix} 1 & 0 \\ 5 & 6z^2+3z+2 \end{bmatrix}.$$

Finally, the $(1,0)$ element is forced to zero by a Type-3 row operation, which is defined by $\begin{bmatrix} 1 & 0 \\ -5 & 1 \end{bmatrix}$. Then, we obtain

$$\begin{bmatrix} 1 & 0 \\ -5 & 1 \end{bmatrix} \begin{bmatrix} 1 & 0 \\ 5 & 6z^2+3z+2 \end{bmatrix} = \begin{bmatrix} 1 & 0 \\ 0 & 6z^2+3z+2 \end{bmatrix}.$$

Thus,

$$\mathbf{S}(z) = \begin{bmatrix} 1 & 0 \\ 0 & 6z^2+3z+2 \end{bmatrix}.$$

Let $\mathbf{E}(z)$ be the product of elementary row operations, i.e.

$$\mathbf{E}(z) = \begin{bmatrix} 1 & 0 \\ -5 & 1 \end{bmatrix} \begin{bmatrix} 1 & -\frac{z}{5} \\ 0 & 1 \end{bmatrix} \begin{bmatrix} 1 & 0 \\ -2(z-1) & 1 \end{bmatrix},$$

or equivalently,

$$\mathbf{E}(z) = \begin{bmatrix} \frac{2z^2-2z+5}{5} & -\frac{z}{5} \\ -2z^2-3 & z+1 \end{bmatrix}.$$

Let $\mathbf{F}(z)$ be the product of elementary column operations, i.e.

$$\mathbf{F}(z) = \begin{bmatrix} 1 & -\frac{3z(2z-1)}{5} \\ 0 & 1 \end{bmatrix},$$

since only one elementary column operation was performed. Therefore,

$$\mathbf{E}(z)\mathbf{A}(z)\mathbf{F}(z) = \mathbf{S}(z).$$

Then, the Smith form decomposition is given by

$$A(z) = U(z)S(z)V(z)$$

where,

$$U(z) = E^{-1}(z) = \begin{bmatrix} \frac{1}{5} & \frac{z^3 - z + 2}{z^2} \\ \frac{2z^2 + 3}{5(z+1)} & \frac{2z^2 - 2z + 5}{z} \end{bmatrix},$$

$$S(z) = \begin{bmatrix} 1 & 0 \\ 0 & 6z^2 + 3z + 2 \end{bmatrix},$$

and

$$V(z) = F^{-1}(z) = \begin{bmatrix} 1 & -\frac{3z(2z-1)}{5} \\ 0 & 1 \end{bmatrix}.$$

4.2.2.3 Theoretical development

Let $H(z)$ be a p x r transfer matrix of rational functions representing a causal Linear Time Invariant system. Assume each element $[H(z)]_{km}$ has been expressed as $[H(z)]_{km} = \frac{P_{km}(z)}{d(z)}$, where the polynomial $d(z)$ is a least common multiple of the denominators of polynomials of $H(z)$. Define a p x r matrix $P(z)$ with elements $P_{km}(z)$ and let

$$P(z) = W(z)\Gamma(z)V(z)$$

be its Smith form decomposition, where $W(z), V(z)$ are unimodular polynomial matrices and the Smith form $\Gamma(z)$ is given by

$$\Gamma(z) = \text{diag}\left(\gamma_0(z), \ldots, \gamma_{r-1}(z), 0, 0, \ldots, 0\right)$$

and γ_i must satisfy $\gamma_i(z)|\gamma_{i+1}(z), i = 0, \ldots, r-2$. Then, the Smith-McMillan decomposition is given by

$$H(z) = W(z)\Lambda(z)V(z)$$

where the Smith-McMillan form is given by

$$\Lambda(z) = \text{diag}\left(\lambda_0(z), \ldots, \lambda_{r-1}(z), 0, 0, \ldots, 0\right)$$

and

$$\lambda_i(z) = \frac{\gamma_i(z)}{d(z)}.$$

Cancelling common factors between $\gamma_i(z)$ and $d(z)$ yields

$$\lambda_i(z) = \frac{\alpha_i(z)}{\beta_i(z)}$$

where $\alpha_i(z)$ and $\beta_i(z)$ are relatively prime polynomials. In view of the divisibility property of the Smith form of $\mathbf{P}(z)$, since $\gamma_i(z)|\gamma_{i+1}(z)$, then $\gamma_{i+1}(z) = c(z)\gamma_i(z)$, and, as a result, $\frac{\alpha_{i+1}(z)}{\beta_{i+1}(z)} = c(z)\frac{\alpha_i(z)}{\beta_i(z)}$. In addition, the polynomials $\alpha_i(z)$ and $\beta_i(z)$ must satisfy $\alpha_i(z)|\alpha_{i+1}(z)$ and $\beta_{i+1}(z)|\beta_i(z), i = 0,\ldots,r-2$.

4.2.3   McMillan degree of a system

Definition 4.2.3.1. *The McMillan degree, μ, of a $p \times r$ causal system $\mathbf{H}(z)$ is the minimum number of delay units (z^{-1} elements) required to implement it, that is*

$$\mu = \deg\,(\mathbf{H}(z)).$$

If the system is noncausal, then the degree is undefined. If $\mathbf{H}(z) = z^{-1}\mathbf{R}$, where $\mathbf{R}$ is an $M \times N$ matrix with rank ρ, then

$$\mathbf{R} = \mathbf{TS}$$

where $\mathbf{T}$ is $M \times \rho$ and $\mathbf{S}$ is $\rho \times N$. Therefore,

$$\mathbf{H}(z) = z^{-1}\mathbf{R} = \mathbf{T}\left[z^{-1}\mathbf{I}_\rho\right]\mathbf{S}.$$

Hence, we can implement the system with ρ delays. So, the system has a McMillan degree $\leq \rho$. As an example, consider

$$\mathbf{H}(z) = z^{-1}\begin{bmatrix} 1 & 2 & 3 \\ 1 & 2 & 3 \\ 1 & 2 & 3 \end{bmatrix}.$$

We can rewrite $\mathbf{H}(z)$ as

$$\mathbf{H}(z) = \begin{bmatrix} 1 \\ 1 \\ 1 \end{bmatrix} z^{-1}\begin{bmatrix} 1 & 2 & 3 \end{bmatrix}.$$

Thus, the system can be implemented with a single delay, illustrating that the McMillan degree of the system is unity.

The Smith-McMillan decomposition provides insight into the determination of the McMillan degree of an $M \times M$ lossless system. The following result is central to the design of lattice structures.

Theorem 4.2.3.1. *Let $\mathbf{H}(z)$ be an $M \times M$ causal lossless system. Then, $\det\,\mathbf{H}(z)$ is stable allpass and*

$$\deg \mathbf{H(z)}\ = \deg\,[\det\mathbf{H(z)}\,].$$

Proof: The stable allpass property of det $\mathbf{H}(z)$ follows from the lossless property of $\mathbf{H}(z)$, that is $\widetilde{\mathbf{H}}(z)\mathbf{H}(z) = c^2\mathbf{I}$. From the Smith-McMillan decomposition,

$$\det \mathbf{H}(z) = a \prod_{i=0}^{M-1} \frac{\alpha_i(z)}{\beta_i(z)}.$$

Since $\alpha_i(z)$ and $\beta_j(z)$ are relatively prime for every i and j, there are no uncancelled factors in the equation. Because of the causality of $\mathbf{H}(z)$, the degree of $\beta_j(z)$ (as a polynomial in z) is at least as large as the degree of $\alpha_j(z)$ (as a polynomial in z). Thus, the degree of the det $\mathbf{H}(z)$ is equal to $\sum_{i=0}^{M-1} \deg \beta_i(z)$, which is the degree of $\mathbf{H}(z)$. ∎

Example 4.2.3.1. The following example illustrates this important result. Consider the following system:

$$\mathbf{H}(z) = \begin{bmatrix} 1 + z^{-1} & 1 - z^{-1} \\ 1 - z^{-1} & 1 + z^{-1} \end{bmatrix}.$$

What is the degree of the system? We will use two methods to determine this degree.

Method 1 (Definition of Smith-McMillan Form)

The Smith-McMillan decomposition of $\mathbf{H}(z)$ is given by

$$\underbrace{\begin{bmatrix} 1 + z^{-1} & 1 - z^{-1} \\ 1 - z^{-1} & 1 + z^{-1} \end{bmatrix}}_{\mathbf{H}(z)} = \underbrace{\begin{bmatrix} -1 & 1 \\ 1 & 1 \end{bmatrix}}_{\mathbf{W}(z)} \underbrace{\begin{bmatrix} z^{-1} & 0 \\ 0 & 1 \end{bmatrix}}_{\Lambda(z)} \underbrace{\begin{bmatrix} -1 & 1 \\ 1 & 1 \end{bmatrix}}_{\mathbf{V}(z)}.$$

So, the Smith-McMillan form of $\mathbf{H}(z)$ is given by

$$\Lambda(z) = \begin{bmatrix} z^{-1} & 0 \\ 0 & 1 \end{bmatrix},$$

or equivalently,

$$\Lambda(z) = \begin{bmatrix} \frac{1}{z} & 0 \\ 0 & 1 \end{bmatrix}.$$

So, $\alpha_0(z) = \alpha_1(z) = 1$, $\beta_0(z) = z$, and $\beta_1(z) = 1$. Therefore, $\mu_0 = \deg \beta_0(z) = 1$ and $\mu_1 = \deg \beta_1(z) = 0$. Hence, by the definition of the Smith-McMillan form, the degree of $\mathbf{H}(z) = \mu_0 + \mu_1 = 1$.

Method 2 (Theorem on McMillan Degree of a Lossless System)

First, we need to determine if $\mathbf{H}(z)$ is paraunitary.

$$\tilde{\mathbf{H}}(z)\mathbf{H}(z) = \begin{bmatrix} 1+z & 1-z \\ 1-z & 1+z \end{bmatrix} \begin{bmatrix} 1+z^{-1} & 1-z^{-1} \\ 1-z^{-1} & 1+z^{-1} \end{bmatrix} = 4\mathbf{I}_2;$$

so, $\mathbf{H}(z)$ is paraunitary and Theorem 4.2.3.1 applies. So compute the determinant of $\mathbf{H}(z)$, that is

$$\det \mathbf{H}(z) = (1+z^{-1})^2 - (1-z^{-1})^2 = 4z^{-1}.$$

So, $\deg [\det \mathbf{H}(\mathbf{z})] = 1$. Therefore, by the theorem on the McMillan degree of a lossless system, $\deg \mathbf{H}(z) = 1$.

4.3 Lattice structures for lossless systems

This section systematically presents lattice structures for lossless systems. Topics included are a fundamental degree-one building block and structures for lossless systems.

4.3.1 Householder factorizations of unitary matrices

Householder factorizations are an important technique for the factorization of unitary matrices. They play an important role in the theory and implementation of lossless systems, because lossless factorizations always involve a unitary matrix.

Definition 4.3.1.1. *Let* $\mathbf{x}$ *be an* M-*dimensional complex-valued vector, such that the ith component of* $\mathbf{x}$ *is given by*

$$x_i = |x_i| \exp(j\theta_i)$$

where θ_i *is real-valued and* $j = (-1)^{1/2}$. *Assume* $\mathbf{x} \neq \mathbf{0}$ *and define*

$$\mathbf{u} = \mathbf{x} + \exp(j\theta_0) \|\mathbf{x}\|_2 \, \mathbf{e}_0$$

where $\mathbf{e}_0$ *is a unit vector in the direction of the zeroth component of* $\mathbf{x}$. *(It is important to note that if* $\mathbf{x}$ *is real-valued, then the definition of* $\mathbf{u}$ *becomes* $\mathbf{u} = \mathbf{x} + \operatorname{sign}(x_0) \|\mathbf{x}\|_2 \, \mathbf{e}_0$.*) The Householder transformation is defined by*

$$\mathbf{H} = \mathbf{I} - \frac{2\mathbf{u}\mathbf{u}^H}{\mathbf{u}^H\mathbf{u}}.$$

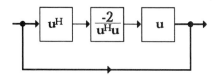

Figure 4.1. Implementation of Householder transformation.

The implementation of a Householder transformation is depicted in Figure 4.1. It is important to note that

$$\mathbf{H}^H = \left(\mathbf{I} - \frac{2\mathbf{u}\mathbf{u}^H}{\mathbf{u}^H\mathbf{u}}\right)^H = \mathbf{I} - \frac{2(\mathbf{u}^H)^H\mathbf{u}^H}{\mathbf{u}^H\mathbf{u}} = \mathbf{I} - \frac{2\mathbf{u}\mathbf{u}^H}{\mathbf{u}^H\mathbf{u}} = \mathbf{H}.$$

Moreover,

$$
\begin{aligned}
\mathbf{H}^H\mathbf{H} &= \left(\mathbf{I} - \tfrac{2\mathbf{u}\mathbf{u}^H}{\mathbf{u}^H\mathbf{u}}\right)\left(\mathbf{I} - \tfrac{2\mathbf{u}\mathbf{u}^H}{\mathbf{u}^H\mathbf{u}}\right) \\
&= \mathbf{I} - 2\left(\tfrac{2\mathbf{u}\mathbf{u}^H}{\mathbf{u}^H\mathbf{u}}\right) + \left(\tfrac{2}{\mathbf{u}^H\mathbf{u}}\right)^2 \mathbf{u}\left(\mathbf{u}^H\mathbf{u}\right)\mathbf{u}^H \\
&= \mathbf{I} - 4\tfrac{\mathbf{u}\mathbf{u}^H}{\mathbf{u}^H\mathbf{u}} + 4\tfrac{\mathbf{u}\mathbf{u}^H}{\mathbf{u}^H\mathbf{u}} \\
&= \mathbf{I}.
\end{aligned}
$$

Hence, $\mathbf{H}$ is a unitary transformation.

In practice, $\mathbf{H}$ never has to be explicitly calculated. This is because $\mathbf{H}\mathbf{x}$ can be predicted *a priori*. But first we need to compute $\mathbf{u}^H\mathbf{x}$ and $\mathbf{u}^H\mathbf{u}$.

$$
\begin{aligned}
\mathbf{u}^H\mathbf{x} &= \left(\mathbf{x} + \exp(j\theta_0)\,\|\mathbf{x}\|_2\,\mathbf{e}_0\right)^H\mathbf{x} \\
&= \mathbf{x}^H\mathbf{x} + \exp(-j\theta_0)\,\|\mathbf{x}\|_2\,\left(\mathbf{e}_0^H\mathbf{x}\right) \\
&= \mathbf{x}^H\mathbf{x} + \exp(-j\theta_0)\,\|\mathbf{x}\|_2 x_0.
\end{aligned}
$$

Since $x_0 = |\, x_0\,|\exp(j\theta_0)$, then

$$\mathbf{u}^H\mathbf{x} = \mathbf{x}^H\mathbf{x} + \|\mathbf{x}\|_2\,|\,x_0\,|\,.$$

So,

$$\mathbf{u}^H\mathbf{u} = \left(\mathbf{x} + \exp(j\theta_0)\,\|\mathbf{x}\|_2\,\mathbf{e}_0\right)^H\left(\mathbf{x} + \exp(j\theta_0)\,\|\mathbf{x}\|_2\,\mathbf{e}_0\right),$$

or equivalently,

$$
\begin{aligned}
\mathbf{u}^H\mathbf{u} &= \mathbf{x}^H\mathbf{x} + \exp(j\theta_0)\|\mathbf{x}\|_2\left(\mathbf{x}^H\mathbf{e}_0\right) + \exp(-j\theta_0)\,\|\mathbf{x}\|_2\,\left(\mathbf{e}_0^H\mathbf{x}\right) \\
&\quad + \left(\exp(-j\theta_0)\,\|\mathbf{x}\|_2\right)\left(\exp(j\theta_0)\,\|\mathbf{x}\|_2\right)\left(\mathbf{e}_0^H\,\mathbf{e}_0\right).
\end{aligned}
$$

Since $\mathbf{x}^H \mathbf{e}_0 = \mid x_0 \mid \exp(-j\theta_0)$, $\mathbf{e}_0^H \mathbf{x} = \mid x_0 \mid \exp(j\theta_0)$, and $\mathbf{e}_0^H\,\mathbf{e}_0 = 1$, then

$$\mathbf{u}^H \mathbf{u} = 2\mathbf{x}^H\mathbf{x} + 2\mid x_0 \mid \|\mathbf{x}\|_2$$

where $\|\mathbf{x}\|_2^2 = \mathbf{x}^H\mathbf{x}$. By combining results, we observe that $\frac{2\mathbf{u}^H\mathbf{x}}{\mathbf{u}^H\mathbf{u}} = 1$. Hence,

$$\mathbf{Hx} = \left(\mathbf{I} - \frac{2\mathbf{u}\mathbf{u}^H}{\mathbf{u}^H\mathbf{u}}\right)\mathbf{x} = \mathbf{x} - \mathbf{u},$$

or equivalently,

$$\mathbf{Hx} = \mathbf{x} - \left(\mathbf{x} + \exp(j\theta_0)\,\|\mathbf{x}\|_2\,\mathbf{e}_0\right).$$

Therefore,

$$\mathbf{Hx} = -\exp(j\theta_0)\,\|\mathbf{x}\|_2\,\mathbf{e}_0.$$

4.3.2   A fundamental degree-one building block

As we saw in the previous section, Householder transformations are given by

$$\mathbf{H} = \mathbf{I} - 2\frac{\mathbf{u}\mathbf{u}^H}{\mathbf{u}^H\mathbf{u}}.$$

Definition 4.3.2.1. *Let $f(z)$ be a function and let $\mathbf{v}_m$ be a vector. Then, the fundamental degree-one structure for lossless systems, denoted $\mathbf{V}_m(z)$, is defined as an extension of the Householder transformation by*

$$\mathbf{V}_m(z) = \mathbf{I} - (1 + f(z))\frac{\mathbf{v}_m\mathbf{v}_m^H}{\mathbf{v}_m^H\mathbf{v}_m}.$$

Let us apply $\mathbf{v}_m$ to $\mathbf{V}_m(z)$, that is,

$$\begin{aligned}
\mathbf{V}_m(z)\mathbf{v}_m &= \left(\mathbf{I} - (1 + f(z))\frac{\mathbf{v}_m\mathbf{v}_m^H}{\mathbf{v}_m^H\mathbf{v}_m}\right)\mathbf{v}_m \\
&= \mathbf{v}_m - \frac{\mathbf{v}_m\left(\mathbf{v}_m^H\mathbf{v}_m\right)}{\mathbf{v}_m^H\mathbf{v}_m} - f(z)\frac{\mathbf{v}_m\left(\mathbf{v}_m^H\mathbf{v}_m\right)}{\mathbf{v}_m^H\mathbf{v}_m} \\
&= -f(z)\mathbf{v}_m.
\end{aligned}$$

So $\mathbf{v}_m$ is an eigenvector with eigenvalue $-f(z)$. Let $\mathbf{u}$ be a vector orthogonal to $\mathbf{v}_m$, that is, $\mathbf{v}_m^H\mathbf{u} = 0$, then

$$\begin{aligned}
\mathbf{V}_m(z)\mathbf{u} &= \left(\mathbf{I} - (1 + f(z))\frac{\mathbf{v}_m\mathbf{v}_m^H}{\mathbf{v}_m^H\mathbf{v}_m}\right)\mathbf{u} \\
&= \mathbf{u} - \frac{\mathbf{v}_m\left(\mathbf{v}_m^H\mathbf{u}\right)}{\mathbf{v}_m^H\mathbf{v}_m} - f(z)\frac{\mathbf{v}_m\left(\mathbf{v}_m^H\mathbf{u}\right)}{\mathbf{v}_m^H\mathbf{v}_m} \\
&= \mathbf{u}.
\end{aligned}$$

So, **u** is an eigenvector with eigenvalue one. Let $\mathbf{u}_k, k = 0, \ldots, M - 1$, be a set of M mutually orthogonal vectors, where $\mathbf{u}_{M-1} = \mathbf{v}_m$. Define an M x M matrix $\mathbf{U} = [\mathbf{u}_0, \ldots, \mathbf{u}_{M-1}]^T$, then we have

$$\mathbf{V}_m(z)\mathbf{U} = \mathbf{U} \begin{bmatrix} \mathbf{I} & 0 \\ 0 & -f(z) \end{bmatrix}.$$

Take the determinant of both sides of the equation

$$\det(\mathbf{V}_m(z)\mathbf{U}) = \det \left(\mathbf{U} \begin{bmatrix} \mathbf{I} & 0 \\ 0 & -f(z) \end{bmatrix} \right).$$

But the determinant of the product of matrices is equal to the product of the determinants of each individual matrix. Hence,

$$\det \mathbf{V}_m(z) \, \det \mathbf{U} = \det \mathbf{U} \, \det \begin{bmatrix} \mathbf{I} & 0 \\ 0 & -f(z) \end{bmatrix},$$

or equivalently,

$$\det \mathbf{V}_m(z) = \det \begin{bmatrix} \mathbf{I} & 0 \\ 0 & -f(z) \end{bmatrix}.$$

Therefore,

$$\det \mathbf{V}_m(z) = -f(z).$$

Since $\mathbf{V}_m(z)$ is lossless, then

$$\begin{aligned} \deg \mathbf{V}_m(z) &= \deg(\det \mathbf{V}_m(z)) \\ &= \deg(-f(z)). \end{aligned}$$

Since $\mathbf{V}_m(z)$ characterizes a degree-one building block, then

$$\deg \mathbf{V}_m(z) = 1.$$

Therefore,

$$\deg(-f(z)) = 1.$$

Let us examine $\mathbf{V}_m(z)$ a little closer. Let

$$\mathbf{P} = \frac{\mathbf{v}_m \mathbf{v}_m^H}{\mathbf{v}_m^H \mathbf{v}_m} \text{ and } \mathbf{Q} = \mathbf{I} - \mathbf{P}.$$

Then,

$$\mathbf{P}^2 = \frac{\mathbf{v}_m (\mathbf{v}_m^H \, \mathbf{v}_m) \mathbf{v}_m^H}{(\mathbf{v}_m^H \mathbf{v}_m)(\mathbf{v}_m^H \mathbf{v}_m)} = \frac{\mathbf{v}_m \mathbf{v}_m^H}{\mathbf{v}_m^H \mathbf{v}_m} = \mathbf{P}$$

and

$$\mathbf{Q}^2 = (\mathbf{I} - \mathbf{P})^2 = \mathbf{I} - 2\mathbf{P} + \mathbf{P}^2 = \mathbf{I} - \mathbf{P} = \mathbf{Q}.$$

Thus, the projection operators $\mathbf{P}$ and $\mathbf{Q}$ satisfy the idempotent property, that is, $\mathbf{P}^2 = \mathbf{P}$ and $\mathbf{Q}^2 = \mathbf{Q}$. In addition,

$$\mathbf{P}^H = \left(\frac{\mathbf{v}_m \mathbf{v}_m^H}{\mathbf{v}_m^H \mathbf{v}_m}\right)^H = \frac{\left(\mathbf{v}_m^H\right)^H \mathbf{v}_m^H}{\mathbf{v}_m^H \mathbf{v}_m} = \frac{\mathbf{v}_m \mathbf{v}_m^H}{\mathbf{v}_m^H \mathbf{v}_m} = \mathbf{P}$$

and

$$\mathbf{Q}^H = (\mathbf{I} - \mathbf{P})^H = \mathbf{I} - \mathbf{P}^H = \mathbf{I} - \mathbf{P} = \mathbf{Q}.$$

Thus, $\mathbf{P}$ and $\mathbf{Q}$ satisfy the Hermitian symmetric property. Since $\mathbf{P}$ and $\mathbf{Q}$ satisfy both the idempotent and symmetric properties, then $\mathbf{P}$ and $\mathbf{Q}$ are orthogonal projections. In addition,

$$\mathbf{PQ} = \mathbf{P}\,(\mathbf{I} - \mathbf{P}) = \mathbf{P} - \mathbf{P}^2 = 0$$

and

$$\mathbf{QP} = (\mathbf{I} - \mathbf{P})\,\mathbf{P} = \mathbf{P} - \mathbf{P}^2 = 0.$$

Consider the function

$$
\begin{aligned}
\widetilde{\mathbf{V}}_m(z)\mathbf{V}_m(z) &= \left(\mathbf{Q}^H - f^*(\tfrac{1}{z^*})\mathbf{P}^H\right)\left(\mathbf{Q} - f(z)\mathbf{P}\right) \\
&= \mathbf{Q}^H\mathbf{Q} - f(z)\mathbf{Q}^H\mathbf{P} - f^*(\tfrac{1}{z^*})\mathbf{P}^H\mathbf{Q} + f^*(\tfrac{1}{z^*})f(z)\mathbf{P}^H\mathbf{P} \\
&= \mathbf{Q}^2 - f(z)\mathbf{QP} - f^*(\tfrac{1}{z^*})\mathbf{PQ} + f^*(\tfrac{1}{z^*})f(z)\mathbf{P}^2 \\
&= \mathbf{Q} + f^*(\tfrac{1}{z^*})f(z)\mathbf{P} \\
&= (\mathbf{Q} + \mathbf{P}) - \left(1 - f^*(\tfrac{1}{z^*})f(z)\right)\mathbf{P} \\
&= \mathbf{I} - \left(1 - f^*(\tfrac{1}{z^*})f(z)\right)\mathbf{P}.
\end{aligned}
$$

For $\mathbf{V}_m(z)$ to be lossless, then

$$1 - f^*(\frac{1}{z^*})f(z) = 0,$$

or equivalently,

$$f(z)f^*(\frac{1}{z^*}) = 1.$$

This implies that poles (and zeros) of $f(z)$ are cancelled by zeros (and poles) of $f^*(\frac{1}{z^*})$. Consequently, $f(z)f^*(\frac{1}{z^*}) = 1$ implies that poles (and zeros) must have zeros (and poles) in conjugate-reciprocal locations. This suggests the following form for $f(z)$, that is,

$$f(z) = \frac{-a_m^* + z^{-1}}{1 - a_m z^{-1}},$$

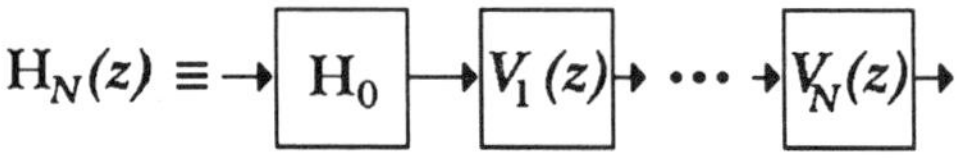

$$H_N(z) \equiv \rightarrow \boxed{H_0} \rightarrow \boxed{V_1(z)} \rightarrow \cdots \rightarrow \boxed{V_N(z)} \rightarrow$$

Figure 4.2. Structure for lossless system.

where a_m is a constant. Therefore, $f(z)$ has a pole at a_m and a zero at $\frac{1}{a_m^*}$. Hence, the fundamental degree-one solution for IIR lossless systems is given by

$$V_m(z) = I - \frac{v_m v_m^H}{v_m^H v_m} + \left(\frac{-a_m^* + z^{-1}}{1 - a_m z^{-1}} \right) \frac{v_m v_m^H}{v_m^H v_m}.$$

For stability, all the poles will be located inside the unit circle so $|a_m| < 1$. When $V_m(z)$ is quantized, it remains unitary, since every occurrence of $v_m v_m^H$ is normalized by $v_m^H v_m$. An interesting special case occurs when $a_m = 0$. Under these conditions, $V_m(z)$ becomes a fundamental building block for FIR lossless systems, that is,

$$V_m(z) = I - \frac{v_m v_m^H}{v_m^H v_m} + z^{-1} \frac{v_m v_m^H}{v_m^H v_m}.$$

Note that the pure delay term z^{-1} is also allpass, since it can be interpreted as a pole at zero and a zero at infinity.

4.3.3 Structures for lossless systems

This section presents lattice structures for lossless systems. Let $H_N(z)$ be an causal lossless system with deg $H_N(z) = N$. Then, we can factorize it as

$$H_N(z) = V_N(z) V_{N-1}(z) \ldots V_1(z) H_0$$

where H_0 is unitary and

$$V_m(z) = I - \frac{v_m v_m^H}{v_m^H v_m} + \left(\frac{-a_m^* + z^{-1}}{1 - a_m z^{-1}} \right) \frac{v_m v_m^H}{v_m^H v_m}.$$

The structure realizing the factorization of $H_N(z)$ is given in Figure 4.2. Sometimes, it is desirable to replace H_0 with its Householder factorization.

Example 4.3.3.1. Factorization of M x 1 FIR Lossless System
Given
$$\mathbf{H}_1(z) = \begin{bmatrix} 0.023 & - & 0.577z^{-1} \\ -0.016 & - & 0.829z^{-1} \end{bmatrix}.$$

Find the M x 1 factorization of $\mathbf{H}_1(z)$.

$\mathbf{H}_1(z)$ can be equivalently written as

$$\mathbf{H}_1(z) = \sum_{j=0}^{1} \mathbf{h}(j)z^{-j} = \underbrace{\begin{bmatrix} 0.023 \\ -0.016 \end{bmatrix}}_{\mathbf{h}(0)} + \underbrace{\begin{bmatrix} -0.577 \\ -0.829 \end{bmatrix}}_{\mathbf{h}(1)} z^{-1}.$$

Since $\mathbf{h}(1)^H \mathbf{h}(0) = 0$, choose $\mathbf{v}_1 = \mathbf{h}(1)$. Then,

$$\widetilde{\mathbf{V}}_1(z) = \mathbf{I} - \frac{\mathbf{h}(1)\mathbf{h}(1)^H}{\mathbf{h}(1)^H \mathbf{h}(1)} + z\frac{\mathbf{h}(1)\mathbf{h}(1)^H}{\mathbf{h}(1)^H \mathbf{h}(1)}.$$

Then,

$$
\begin{aligned}
\mathbf{H}_0 &= \widetilde{\mathbf{V}}_1(z)\mathbf{H}_1(z) \\
&= \left[\mathbf{I} - \frac{\mathbf{h}(1)\mathbf{h}(1)^H}{\mathbf{h}(1)^H\mathbf{h}(1)} + z\frac{\mathbf{h}(1)\mathbf{h}(1)^H}{\mathbf{h}(1)^H\mathbf{h}(1)}\right]\left[\mathbf{h}(0) + \mathbf{h}(1)z^{-1}\right] \\
&= \mathbf{h}(0) + \mathbf{h}(1)z^{-1} - \mathbf{h}(1)z^{-1} + \mathbf{h}(1) \\
&= \mathbf{h}(0) + \mathbf{h}(1).
\end{aligned}
$$

Hence,

$$\mathbf{H}_0 = \begin{bmatrix} -0.554 \\ -0.845 \end{bmatrix}.$$

For constant vector, perform a Householder transformation, that is,

$$\mathbf{U}_0 = \mathbf{I} - \frac{2\mathbf{u}_0\mathbf{u}_0^H}{\mathbf{u}_0^H\mathbf{u}_0}$$

where

$$\mathbf{u}_0 = \begin{bmatrix} \text{sign}\,(\mathbf{H}_0)_0\|\mathbf{H}_0\|_2 & + & (\mathbf{H}_0)_0 \\ & & (\mathbf{H}_0)_1 \end{bmatrix} = \begin{bmatrix} -1.574 \\ -0.845 \end{bmatrix}.$$

Therefore,

$$\mathbf{H}_1(z) = \mathbf{V}_1(z)\mathbf{U}_0.$$

As such, the filter $\mathbf{H}_1(z)$ can be realized by two stages $\mathbf{V}_1(z)$ and $\mathbf{U}_0$. This filter is depicted in Figure 4.3.

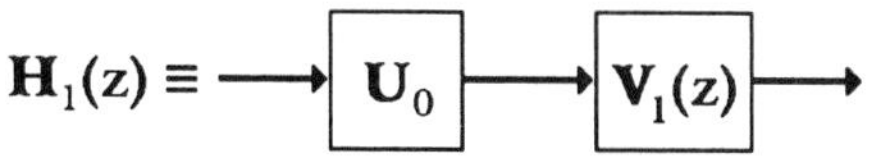

$$\mathbf{H_1(z)} \equiv \longrightarrow \boxed{\mathbf{U_0}} \longrightarrow \boxed{\mathbf{V_1(z)}} \longrightarrow$$

Figure 4.3. Two-stage filter.

Example 4.3.3.2. Factorization of M x M FIR Lossless System

Given some positive integer k, then the system $\mathbf{H}_k(z)$ is given by

$$\mathbf{H}_k(z) = \begin{bmatrix} 1 + z^{-k} & 1 - z^{-k} \\ 1 - z^{-k} & 1 + z^{-k} \end{bmatrix}.$$

Find the M x M factorization of $\mathbf{H}_k(z)$.

By inspection, $\mathbf{H}_k(z)$ is FIR. Is $\mathbf{H}_k(z)$ paraunitary?

$$\widetilde{\mathbf{H}}_k(z)\mathbf{H}_k(z) = \begin{bmatrix} 1 + z^k & 1 - z^k \\ 1 - z^k & 1 + z^k \end{bmatrix} \begin{bmatrix} 1 + z^{-k} & 1 - z^{-k} \\ 1 - z^{-k} & 1 + z^{-k} \end{bmatrix} = 4\,\mathbf{I}_2.$$

Hence, $\mathbf{H}_k(z)$ is stable and paraunitary. Therefore,

$$\deg \mathbf{H}_k(z) = \deg(\det \mathbf{H}_k(z)).$$

So,

$$\deg \mathbf{H}_k(z) = \deg[(1 + z^{-k})^2 - (1 - z^{-k})^2] = \deg[4z^{-k}] = k.$$

The impulse response matrix of $\mathbf{H}_k(z)$ is given by

$$\mathbf{H}_k(z) = \underbrace{\begin{bmatrix} 1 & 1 \\ 1 & 1 \end{bmatrix}}_{\mathbf{h}(0)} + \underbrace{\begin{bmatrix} 1 & -1 \\ -1 & 1 \end{bmatrix}}_{\mathbf{h}(k)} z^{-k}.$$

Choose $\mathbf{v}_k$ such that $\mathbf{v}_k^H \mathbf{h}(0) = \mathbf{0}$. So, let $\mathbf{v}_k = \begin{bmatrix} 1 \\ -1 \end{bmatrix}$. Then,

$$\mathbf{H}_k(z) = \mathbf{V}_k(z)\mathbf{H}_{k-1}(z).$$

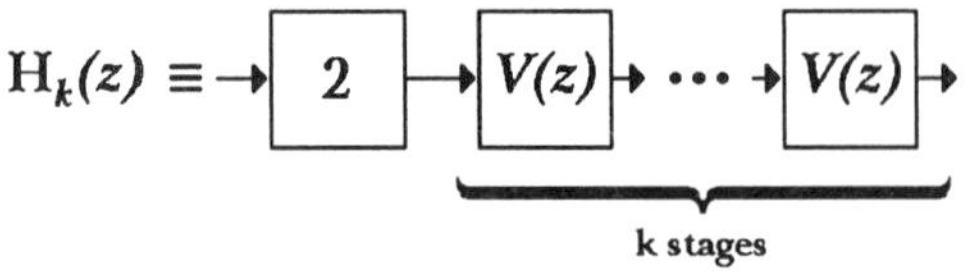

Figure 4.4. K-stage filter.

Hence,

$$\mathbf{H}_{k-1} = \widetilde{\mathbf{V}}_k(z)\mathbf{H}_k(z)$$

$$= \left(\begin{bmatrix} 1 & 0 \\ 0 & 1 \end{bmatrix} - \frac{1}{2} \begin{bmatrix} 1 & -1 \\ -1 & 1 \end{bmatrix} + z\left(\frac{1}{2}\right) \begin{bmatrix} 1 & -1 \\ -1 & 1 \end{bmatrix} \right)$$

$$x\left(\begin{bmatrix} 1 & 1 \\ 1 & 1 \end{bmatrix} + z^{-k} \begin{bmatrix} 1 & -1 \\ -1 & 1 \end{bmatrix} \right)$$

$$= \begin{bmatrix} 1 & 1 \\ 1 & 1 \end{bmatrix} + z^{-k} \begin{bmatrix} 1 & -1 \\ -1 & 1 \end{bmatrix} - z^{-k} \begin{bmatrix} 1 & -1 \\ -1 & 1 \end{bmatrix}$$

$$+ z^{-(k-1)} \begin{bmatrix} 1 & -1 \\ -1 & 1 \end{bmatrix}$$

$$= \begin{bmatrix} 1 + z^{-(k-1)} & 1 - z^{-(k-1)} \\ 1 - z^{-(k-1)} & 1 + z^{-(k-1)} \end{bmatrix}.$$

Then,

$$\mathbf{V}(z) = \mathbf{V}_1(z) = \cdots = \mathbf{V}_k(z)$$

will be the degree-one building block. Then,

$$\mathbf{H}_0 = \widetilde{\mathbf{V}}_1(z)\mathbf{H}_1(z) = \begin{bmatrix} 2 & 0 \\ 0 & 2 \end{bmatrix} = 2\mathbf{I}_2.$$

Therefore, the filter can be realized by k stages $\mathbf{V}(z)$ and a constant scaling factor of 2. This filter is depicted in Figure 4.4.

4.4 Problems

1. Let $G(z)$ be given by the following:

$$G(z) = \begin{bmatrix} 1 + z^{-1} & 2 + z^{-1} \\ 2 + z^{-1} & 3 + z^{-1} \end{bmatrix}.$$

Express $G(z)$ as a sum of impulse response matrices. Is $G(z)$ unimodular?

2. Let B be given by the following:

$$B = \begin{bmatrix} 1 + ja & 1 + ja \\ 1 - ja & -1 + ja \end{bmatrix},$$

where a is a real-valued constant. Use a Householder transformation to make B an upper triangular matrix.

3. Let $H(z)$ be a system transfer function that is given by the following:

$$H(z) = \begin{bmatrix} 1 & -cz^{-k} \\ c & z^{-k} \end{bmatrix},$$

where c is a real-valued constant. Is $H(z)$ paraunitary? Draw the cascaded structure to implement $H(z)$.

4. Let $H(z)$ be given by the following

$$H(z) = \begin{bmatrix} 1 + 2z^{-1} - z^{-2} & 1 + z^{-2} \\ 1 + z^{-2} & 1 - 2z^{-1} - z^{-2} \end{bmatrix}.$$

Is $H(z)$ lossless? What is the degree of $H(z)$? Obtain a cascaded structure to implement $H(z)$.

5. Find the Smith-McMillan decomposition of the $H(z)$ matrix, which was defined in the last problem. Then, using this result, determine the degree of the system.

Chapter 5

Wavelet Signal Processing

5.1 Introduction

This chapter introduces the notion of fundamentals of wavelet signal processing. Morlet *et al.*[36] was the first paper to discuss the idea of a wavelet, although at this point in time it was a rather empirical idea. Later, Daubechies[12] showed that it was possible to develop a theory and an associated algorithm for the generation of compactly supported orthonormal wavelets. Meanwhile, Mallat[27, 28] developed the notions of multiresolution analysis; and, recently, Shapiro[42] has developed an efficient scheme for encoding them. Results on biorthogonal wavelets can be found in Cohen *et al.*[8].

Many of the concepts presented in this chapter are also discussed in other multirate and wavelet texts, including the engineering-oriented texts by Chan[5], Fliege[19], Strang and Nguyen[46], Vaidyanathan[49], Vetterli and Kovačević[54], Wornell[57] and the mathematically-oriented texts by Chui[7], Daubechies[13], Holschneider[22], and Meyer[33, 34]. Good reference texts for background material on vector spaces include Riesz and Sz.-Nagy[39] and Schilling and Lee[41].

Section 5.2 introduces the wavelet transform. Section 5.3 presents multiresolution analysis for both orthogonal and biorthogonal wavelets.

5.2 Wavelet transform

The name *wavelet* comes from the requirement that a function should integrate to zero, *waving* above and below the axis. The diminutive connotation of wavelet suggests the function has to be well localized. At this point, some people might ask, why not use traditional Fourier methods? Fourier basis functions are localized in frequency, but not in time. So, Fourier analysis is the ideal tool for the efficient representation of very smooth,

167

stationary signals. Wavelet basis functions are localized in time and frequency. So, wavelet analysis is an ideal tool for representing signals that contain discontinuities (in the signal or its derivatives) or for signals that are not stationary.

5.2.1   Definition and properties

Definition 5.2.1.1. *The wavelet transform of $g(t)$ with respect to wavelet $\psi(t)$ is defined by*

$$\mathrm{WT}\{g;\ a,b\} = \frac{1}{\sqrt{|a|}} \int_{-\infty}^{\infty} g(t)\ \psi^*\left(\frac{t-b}{a}\right)\ dt$$

where, $a \neq 0$ and b are called the scale and translation parameters, respectively. Furthermore, the Fourier transform of wavelet $\psi(t)$, denoted $\Psi(f)$, must satisfy the following admissibility condition:

$$C_\psi = \int_{-\infty}^{\infty} \frac{|\Psi(f)|^2}{f}\ df < \infty,$$

which shows that $\psi(t)$ has to oscillate and decay. The inverse of the continuous wavelet transform is given by

$$g(t) = \frac{1}{C_\psi} \int_{-\infty}^{\infty} \int_{-\infty}^{\infty} \mathrm{WT}\{g;\ a,b\}\ \psi\left(\frac{t-b}{a}\right) \frac{da\ db}{a^2}.$$

It is often useful to think of functions and their wavelet transforms as occupying two domains. Then, the following properties show the correspondence between operations performed in one domain with operations in the other.

5.2.1.1   Linearity property

Theorem 5.2.1.1. *If the wavelet transforms of g and h exist and α and β are scalars, then*

$$\mathrm{WT}\{\alpha g + \beta h;\ a,b\} = \alpha \mathrm{WT}\{g;\ a,b\} + \beta \mathrm{WT}\{h;\ a,b\}.$$

Proof: By definition,

$$\mathrm{WT}\{\alpha g + \beta h;\ a,b\} = \frac{1}{\sqrt{|a|}} \int_{-\infty}^{\infty} \{\alpha g(t) + \beta h(t)\}\ \psi^*\left(\frac{t-b}{a}\right)\ dt.$$

By the linearity of integration,

$$
\begin{aligned}
\mathrm{WT}\{\alpha g + \beta h;\ a,b\} = \ & \frac{\alpha}{\sqrt{|a|}} \int_{-\infty}^{\infty} g(t)\ \psi^*\left(\frac{t-b}{a}\right)\ dt \\
& + \frac{\beta}{\sqrt{|a|}} \int_{-\infty}^{\infty} h(t)\ \psi^*\left(\frac{t-b}{a}\right)\ dt.
\end{aligned}
$$

Hence,

$$\text{WT}\{\alpha g + \beta h;\ a,b\} = \alpha \,\text{WT}\{g;\ a,b\} + \beta \,\text{WT}\{h;\ a,b\}.$$

∎

5.2.1.2 Similarity theorem

Theorem 5.2.1.2. *If the wavelet transform of f exists, then the wavelet transform of $f(\alpha \bullet)$, for a constant α, is given by*

$$\text{WT}\{f(\alpha \bullet); a,b\} = \frac{1}{\sqrt{|\alpha|}}\text{WT}\{f; \alpha a, \alpha b\}.$$

Proof: By definition,

$$\text{WT}\{f(\alpha \bullet); a,b\} = \frac{1}{\sqrt{|a|}}\int_{-\infty}^{\infty} f(\alpha t)\,\psi^*\!\left(\frac{t-b}{a}\right)\,dt.$$

Let $u = \alpha t$. Then,

$$\text{WT}\{f(\alpha \bullet); a,b\} = \frac{1}{|\alpha|\sqrt{|a|}}\int_{-\infty}^{\infty} f(u)\,\psi^*\!\left(\frac{\frac{u}{\alpha}-b}{a}\right)\,du,$$

or equivalently,

$$\text{WT}\{f(\alpha \bullet); a,b\} = \frac{1}{\sqrt{|\alpha|}\sqrt{|\alpha a|}}\int_{-\infty}^{\infty} f(u)\,\psi^*\!\left(\frac{u-\alpha b}{\alpha a}\right)du.$$

Hence,

$$\text{WT}\{f(\alpha \bullet); a,b\} = \frac{1}{\sqrt{|\alpha|}}\text{WT}\{f; \alpha a, \alpha b\}.$$

∎

5.2.1.3 Shift theorem

Theorem 5.2.1.3. *If the wavelet transformof f exists, then the wavelet transform of $f(\bullet - \alpha)$, for a constant α, is given by*

$$\text{WT}\{f(\bullet - \alpha); a,b\} = \text{WT}\{f; a, b-\alpha\}.$$

Proof: By definition,

$$\text{WT}\{f(\bullet - \alpha); a, b\} = \frac{1}{\sqrt{|a|}} \int_{-\infty}^{\infty} f(t - \alpha)\, \psi^*\Big(\frac{t - b}{a}\Big)\, dt.$$

Let $u = t - \alpha$. Then,

$$\text{WT}\{f(\bullet - \alpha); a, b\} = \frac{1}{\sqrt{|a|}} \int_{-\infty}^{\infty} f(u)\, \psi^*\left(\frac{u - (b - \alpha)}{a}\right)\, du,$$

or equivalently,

$$\text{WT}\{f(\bullet - \alpha); a, b\} = \text{WT}\{f; a, b - \alpha\}.$$

$\blacksquare$

5.2.1.4   Differentiation theorem

Theorem 5.2.1.4. *If the wavelet transform of f exists and if f' exists, then the wavelet transform of f' is given by*

$$\text{WT}\{f'; a, b\} = \frac{\partial}{\partial b} \text{WT}\{f; a, b\}.$$

Proof: By definition,

$$\text{WT}\{f'; a, b\} = \frac{1}{\sqrt{|a|}} \int_{-\infty}^{\infty} f'(t)\, \psi^*\Big(\frac{t - b}{a}\Big)\, dt.$$

But for every real t,

$$f'(t) = \lim_{\epsilon \to 0} \frac{f(t + \epsilon) - f(t)}{\epsilon}.$$

Then, using Lebesque's Dominated Convergence Theorem, we obtain

$$\text{WT}\{f'; a, b\} = \lim_{\epsilon \to 0} \frac{1}{\sqrt{|a|}} \int_{-\infty}^{\infty} \left(\frac{f(t + \epsilon) - f(t)}{\epsilon}\right) \psi^*\Big(\frac{t - b}{a}\Big)\, dt.$$

Using the Linearity Theorem, we obtain

$$\text{WT}\{f'; a, b\} = \lim_{\epsilon \to 0} \frac{1}{\epsilon}\Big(\frac{1}{\sqrt{|a|}} \int_{-\infty}^{\infty} f(t + \epsilon)\, \psi^*(\tfrac{t-b}{a})\, dt$$
$$- \frac{1}{\sqrt{|a|}} \int_{-\infty}^{\infty} f(t)\, \psi^*(\tfrac{t-b}{a})\, dt\Big).$$

Using the Shift Theorem, we obtain

$$\text{WT}\{f'; a, b\} = \lim_{\epsilon \to 0} \frac{\text{WT}\{f; a, b + \epsilon\} - \text{WT}\{f; a, b\}}{\epsilon}.$$

Therefore,

$$\text{WT}\{f'; a, b\} = \frac{\partial}{\partial b} \text{WT}\{f; a, b\}.$$

$\blacksquare$

5.2.1.5   Convolution theorem

Theorem 5.2.1.5. *If the wavelet transform of* g *exists and if* $f \star g$ *exists, then the wavelet transform of* $f \star g$ *is given by*

$$\text{WT}\{f \star g; a, b\} = f \overset{b}{\star} \text{WT}\{g; a, b\},$$

where $\overset{b}{\star}$ *denotes convolution with respect to the* b *variable.*

Proof: By definition,

$$\text{WT}\{f \star g; a, b\} = \frac{1}{\sqrt{|a|}} \int_{-\infty}^{\infty} \left[\int_{-\infty}^{\infty} f(u)g(t-u)du\right] \psi^*(\frac{t-b}{a})\, dt,$$

or equivalently,

$$\text{WT}\{f \star g; a, b\} = \int_{-\infty}^{\infty} f(u) \left[\frac{1}{\sqrt{|a|}} \int_{-\infty}^{\infty} g(t-u)\psi^*(\frac{t-b}{a})dt\right] du.$$

Using the Shift Theorem we obtain,

$$\text{WT}\{f \star g; a, b\} = \int_{-\infty}^{\infty} f(u)\text{WT}\{g; a, b-u\}du,$$

or equivalently,

$$\text{WT}\{f \star g; a, b\} = f \overset{b}{\star} \text{WT}\{g; a, b\}.$$

∎

    The following subsection illustrates the role that wavelets naturally play in radar signal processing.

5.2.2   Radar signal processing and wavelets

The problem consists of estimating the location and velocity of some target in a radar application. The estimation procedure can be described by the following. Suppose $x(t)$ is a known emitting signal. In the presence of a target, this signal $x(t)$ will return to the source as the received signal $h(t)$ with a delay τ, due to the target's location and a Doppler effect distortion, due to the target's velocity. If $x(t)$ is a narrow-band signal, then the Doppler effect amounts to a single frequency shift f_0. The characteristics of the target can be determined by maximizing the cross-correlation function. This estimator is called the *Narrow-Band Ambiguity Function*, that is,

$$\int_{-\infty}^{\infty} x(t)\, h(t-\tau)\, \exp\left(-j2\pi f_0 t\right)\, dt$$

which is simply the Short Time Fourier Transform (STFT) with shift τ about frequency f_0.

However, for wide-band signals, the Doppler frequency shift varies the signal spectrum, causing a stretching or compression of the emitted signal. The resulting estimator is called the *Wide-Band Ambiguity Function*, that is

$$(|\,a\,|)^{-1/2} \int_{-\infty}^{\infty} x(t)\, h\left(\frac{t-\tau}{a}\right) dt$$

which is continuous wavelet transform (CWT) about a point τ at a scale given by a. Thus, the wavelet transform is an operation that determines the similarity or cross correlation between the emitted signal $x(t)$ and the received signal, the wavelet $h\left(\frac{t-\tau}{a}\right)$ at scale a and shift τ.

5.3   Multiresolution analysis

As we have observed, the wavelet transform maps a function of one dimension into a two-dimensional picture. Using this idea, a mathematical microscope known as *multiresolution analysis* will permit us to look at the details that are added as we go from one scale to another. In this way, multiresolution analysis provides a mathematical framework to conceptualize problems linked to the wavelet decomposition of signals.

5.3.1   Definition, properties, and implementation

Before proceeding with the definition of multiresolution analysis, we will first need to discuss four topics of great importance for the study of multiresolution analysis: inner products, biorthogonality, Riesz basis, and direct sums.

5.3.1.1   Inner products

Definition 5.3.1.1. *The inner product on a vector space* **V** *is a mapping that assigns a scalar to every pair of elements* **s, t** $\in$ **V**. *The inner product is denoted* $\langle$**s, t**$\rangle$ *and it has the following properties:*

1. $\langle$**s, t**$\rangle^* = \langle$**t, s**$\rangle$;
2. $\langle a$**s**$+b$**t, u**$\rangle = a\langle$**s, u**$\rangle + b\langle$**t, u**$\rangle$;
3. $\langle$**s, s**$\rangle \geq 0$;
4. $\langle$**s, s**$\rangle = 0 \Leftrightarrow$ **s** $= 0$.

A vector space **V** with an inner product is called an inner product space. The operation of performing the inner product is defined by the following theorem.

Theorem 5.3.1.1. Let $\mathbf{V}$ be a finite-dimensional inner product space. Suppose $\{\mathbf{u}_1, \ldots, \mathbf{u}_n\}$ and $\{\mathbf{x}_1, \ldots, \mathbf{x}_n\}$ are two basis sets for $\mathbf{V}$. Let vectors $\mathbf{s}$ and $\mathbf{t}$ be given in terms of the basis set $\{\mathbf{u}_1, \ldots, \mathbf{u}_n\}$ and $\{\mathbf{x}_1, \ldots, \mathbf{x}_n\}$, respectively. That is,

$$\mathbf{s} = \sum_{k=1}^{n} s_k \mathbf{u}_k \text{ and } \mathbf{t} = \sum_{k=1}^{n} t_k \mathbf{x}_k.$$

Let the weighting matrix $\mathbf{W}$ be a defined by

$$[\mathbf{W}]_{kj} = \langle \mathbf{u}_j, \mathbf{x}_k \rangle.$$

Then, the inner product is given by

$$\langle \mathbf{s}, \mathbf{t} \rangle = \mathbf{t}^H \mathbf{W} \mathbf{s}.$$

Proof: Consider the inner product of $\mathbf{s}$ and $\mathbf{t}$ by first substituting their respective series expansions, that is,

$$\langle \mathbf{s}, \mathbf{t} \rangle = \left\langle \sum_{k=1}^{n} s_k \mathbf{u}_k, \sum_{j=1}^{n} t_j \mathbf{x}_j \right\rangle.$$

Utilizing the linearity property of the inner product yields

$$\langle \mathbf{s}, \mathbf{t} \rangle = \sum_{k=1}^{n} s_k \sum_{j=1}^{n} t_j^* \langle \mathbf{u}_k, \mathbf{x}_j \rangle.$$

By definition, $[\mathbf{W}]_{kj} = \langle \mathbf{u}_j, \mathbf{x}_k \rangle$. Hence,

$$\langle \mathbf{s}, \mathbf{t} \rangle = \sum_{j=1}^{n} t_j^* \sum_{k=1}^{n} [\mathbf{W}]_{kj} \, s_k,$$

or equivalently,

$$\langle \mathbf{s}, \mathbf{t} \rangle = \sum_{j=1}^{n} t_j^* \, (\mathbf{W}\mathbf{s})_j.$$

Therefore,

$$\langle \mathbf{s}, \mathbf{t} \rangle = \mathbf{t}^H \mathbf{W} \mathbf{s}.$$

∎

5.3.1.2   Biorthogonality

Definition 5.3.1.2. *Two linearly independent sets of vectors* $S = \{u_1, \ldots,$ $u_n\}$ *and* $T = \{x_1, \ldots, x_n\}$ *form biorthogonal sets if*

$$\langle u_j, x_k \rangle = 0 \text{ for all } j \text{ and } k \text{ where } j \neq k,$$

and

$$\langle u_j, x_j \rangle \neq 0 \text{ for all } j,$$

or equivalently, W *is a diagonal matrix.*
Furthermore, S *and* T *are called biorthonormal sets if*

$$\langle u_j, x_k \rangle = \delta_{j,k},$$

or equivalently, W *is an identity matrix. Moreover, any set of biorthogonal vectors can be made biorthonormal. If* $S = T$, *then biorthogonal sets become orthogonal sets and biorthonormal sets become orthonormal sets.*

The following theorem deals with transformations that preserve all the information of the original signal.

Theorem 5.3.1.2. *Let* $\{x_k\}$ *and* $\{u_k\}$ *be biorthonormal bases of an inner product space* V. *Then,* $y \in V$ *can be expressed as*

$$y = \sum_k \langle y, x_k \rangle \, u_k.$$

Proof: Since $y \in V$ and since $\{x_k\}$ and $\{u_k\}$ are complete biorthonormal sets of nonzero vectors in an inner product space V, then y can be expressed as

$$y = \sum_k c_k \, u_k.$$

Taking the inner product of both sides of this equation with x_m yields

$$\langle y, x_m \rangle = \left\langle \sum_k c_k \, u_k, x_m \right\rangle,$$

or by linearity,

$$\langle y, x_m \rangle = \sum_k c_k \, \langle u_k, x_m \rangle.$$

Since u_k and x_m are biorthonormal vectors, then

$$\langle y, x_m \rangle = \sum_k c_k \delta_{k,m}.$$

Therefore, the right-hand side is nonzero only if $k = m$. Hence,

$$\langle \mathbf{y}, \mathbf{x}_k \rangle = c_k.$$

Hence,

$$\mathbf{y} = \sum_k \langle \mathbf{y}, \mathbf{x}_k \rangle \, \mathbf{u}_k.$$

∎

The following theorem shows the interrelationship between paraunitary matrices and biorthogonality.

Theorem 5.3.1.3. *The filter* $\mathbf{H}(z)$ *satisfies the normalized paraunitary property*

$$\widetilde{\mathbf{H}}(z)\mathbf{H}(z) = \mathbf{I}_M$$

if and only if its individual filters satisfy the biorthonormal condition

$$\left\langle \widetilde{\mathbf{h}}_j(z), \mathbf{h}_k(z) \right\rangle = \delta_{j,k}$$

where,

$$\mathbf{h}_k(z) = \left[H_k(z), \ H_k(zW_M), \dots, H_k(zW_M^{M-1}) \right]^T.$$

Proof: (Case 1: Assume $\widetilde{\mathbf{H}}(z)\mathbf{H}(z) = \mathbf{I}_M$.) Then, the (j, k)th entry in the matrix $\widetilde{\mathbf{H}}(z)\mathbf{H}(z)$ is given by

$$\left[\widetilde{\mathbf{H}}(z)\mathbf{H}(z) \right]_{j,k} = \left\langle \widetilde{\mathbf{h}}_j(z), \mathbf{h}_k(z) \right\rangle.$$

Since $\widetilde{\mathbf{H}}(z)\mathbf{H}(z) = \mathbf{I}_M$, then

$$\left\langle \widetilde{\mathbf{h}}_j(z), \mathbf{h}_k(z) \right\rangle = \delta_{j,k}.$$

(Case 2: Assume $\left(\left\langle \widetilde{\mathbf{h}}_j(z), \mathbf{h}_k(z) \right\rangle = \delta_{j,k}.\right)$ If this equation is interpreted as the (j, k)th entry of a matrix, then this equation can be rewritten as

$$\left[\widetilde{\mathbf{H}}(z)\mathbf{H}(z) \right]_{j,k} = \delta_{j,k}.$$

Then, the corresponding matrix equation is given by

$$\widetilde{\mathbf{H}}(z)\mathbf{H}(z) = \mathbf{I}_M.$$

∎

5.3.1.3   Riesz basis

Definition 5.3.1.3. *Let* **V** *be an inner product space. Let* $\| \bullet \|$ *denote the norm induced by the inner product, that is,*

$$\|\mathbf{s}\|^2 = \langle \mathbf{s}, \mathbf{s} \rangle .$$

The set $\{\mathbf{f}_k\} \subset \mathbf{V}$ *is called a Riesz Basis if every element* $\mathbf{s} \in \mathbf{V}$ *of the space can be written as*

$$\mathbf{s} = \sum_k c_k \, \mathbf{f}_k$$

for some choice of scalars $\{c_k\}$ *and if positive constants* A *and* B *exist such that*

$$A\|\mathbf{s}\|^2 \leq \sum_k |c_k|^2 \leq B\|\mathbf{s}\|^2 .$$

Riesz basis are also known as a stable basis or unconditional basis. If the Reisz basis is an orthogonal basis, then $A = B = 1$.

Consider two biorthonormal sets $\mathbf{S} = \{\mathbf{u}_1, \ldots, \mathbf{u}_n\}$ and $\mathbf{T} = \{\mathbf{x}_1, \ldots, \mathbf{x}_n\}$, that is,

$$\langle \mathbf{u}_j, \mathbf{x}_k \rangle = \delta_{j,k}.$$

Then, $\mathbf{S}$ and $\mathbf{T}$ are biorthonormal (Riesz) basis if there exist positive constants A, B, C, D, such that for all vectors $\mathbf{y}$

$$A\|\mathbf{y}\|^2 \leq \sum_k |\langle \mathbf{u}_k, \mathbf{y} \rangle|^2 \leq B\|\mathbf{y}\|^2$$

and

$$C\|\mathbf{y}\|^2 \leq \sum_k |\langle \mathbf{x}_k, \mathbf{y} \rangle|^2 \leq D\|\mathbf{y}\|^2.$$

5.3.1.4   Direct sums

Now, we will define the concept of a direct sum and, subsequently, we will prove the projection theorem.

Definition 5.3.1.4. *Let* **S** *be a subspace and let* **U** *and* **V** *be subspaces of* **S**. *Then,* **S** *can be decomposed into a direct sum of* **U** *and* **V**, *that is*

$$\mathbf{S} = \mathbf{U} \oplus \mathbf{V},$$

if the following conditions are satisfied:

1. $\mathbf{U} \cap \mathbf{V} = \mathbf{0}$
2. **S** *is generated by* $\mathbf{U} \cup \mathbf{V}$.

We are now in a position to decompose a vector into its projections onto subspaces. To introduce this concept, we begin by identifying suitable subspaces.

Definition 5.3.1.5. *The orthogonal complement of a subset* $\mathbf{U}$ *of an inner product space* $\mathbf{V}$ *is denoted* $\mathbf{U}^\perp$ *and is defined by*

$$\mathbf{U}^\perp = \{\mathbf{x} \in \mathbf{V} \mid \forall \, \mathbf{y} \in \mathbf{U} \Rightarrow \langle \mathbf{x}, \mathbf{y} \rangle = 0\}.$$

Thus, $\mathbf{U}^\perp$ will always be a subspace of $\mathbf{V}$ and that subspace will be closed. Moreover, if $\mathbf{U}$ is a closed subspace, then $\mathbf{U}^{\perp\perp} = \mathbf{U}$.

Theorem 5.3.1.4. (Projection Theorem) *Let* $\mathbf{U}$ *be a closed subspace of a complete inner product space* $\mathbf{V}$. *Then,* $\mathbf{V}$ *can be decomposed into a direct sum of* $\mathbf{U}$ *and* $\mathbf{U}^\perp$, *that is,*

$$\mathbf{V} = \mathbf{U} \oplus \mathbf{U}^\perp.$$

Proof: Let $\mathbf{Y} = \{\mathbf{y}_1, \ldots, \mathbf{y}_n\}$ and $\mathbf{S} = \{\mathbf{s}_1, \ldots, \mathbf{s}_n\}$ be an biorthonormal basis for $\mathbf{U}$. Let $\mathbf{x} \in \mathbf{V}$. Then, $\mathbf{x}_0$ and $\mathbf{x}_1$ are defined by

$$\mathbf{x}_0 = \sum_{k=1}^{n} \langle \mathbf{x}, \mathbf{y}_k \rangle \mathbf{s}_k$$

and

$$\mathbf{x}_1 = \mathbf{x} - \mathbf{x}_0.$$

Clearly, $\mathbf{x} = \mathbf{x}_0 + \mathbf{x}_1$. Furthermore, $\mathbf{x}_0 \in \mathbf{U}$ since $\mathbf{x}_0$ is a linear combination of the basis elements $\mathbf{Y}$. Next, we must demonstrate that $\mathbf{x}^\perp \in \mathbf{U}^\perp$ and that the decomposition of $\mathbf{x}$ is unique. Let us consider the inner product between $\mathbf{x}_1$ and $\mathbf{y}_k \in \mathbf{Y}$, that is,

$$\langle \mathbf{x}_1, \mathbf{y}_k \rangle = \langle \mathbf{x} - \mathbf{x}_0, \mathbf{y}_k \rangle,$$

or equivalently,

$$\langle \mathbf{x}_1, \mathbf{y}_k \rangle = \langle \mathbf{x}, \mathbf{y}_k \rangle - \langle \mathbf{x}_0, \mathbf{y}_k \rangle.$$

Now replacing $\mathbf{x}_0$ with its series expansion yields

$$\langle \mathbf{x}_1, \mathbf{y}_k \rangle = \langle \mathbf{x}, \mathbf{y}_k \rangle - \left\langle \sum_{j=1}^{n} \langle \mathbf{x}, \mathbf{y}_j \rangle \mathbf{s}_j, \mathbf{y}_k \right\rangle,$$

or equivalently,

$$\langle \mathbf{x}_1, \mathbf{y}_k \rangle = \langle \mathbf{x}, \mathbf{y}_k \rangle - \sum_{j=1}^{n} \langle \mathbf{x}, \mathbf{y}_j \rangle \langle \mathbf{s}_j, \mathbf{y}_k \rangle.$$

Since $\mathbf{Y}$ and $\mathbf{S}$ form a biorthonormal set, $\langle \mathbf{s}_j, \mathbf{y}_k \rangle = \delta_{j,k}$. Hence,

$$\langle \mathbf{x}_1, \mathbf{y}_k \rangle = \langle \mathbf{x}, \mathbf{y}_k \rangle - \langle \mathbf{x}, \mathbf{y}_k \rangle = 0.$$

Thus, $\mathbf{x}_1$ is orthogonal to each basis vector in $\mathbf{Y}$. Since each vector in $\mathbf{U}$ is a linear combination of the basis vectors, it follows that $\mathbf{x}_1$ is orthogonal to every vector in $\mathbf{U}$. Hence, $\mathbf{x}_1 \in \mathbf{U}^{\perp}$.

To show the decomposition of $\mathbf{x}$ is unique, let us assume that $\mathbf{x}$ is not unique and show that this results in a contradiction. Let's assume $\mathbf{x} = \mathbf{x}_0 + \mathbf{x}_1$ and $\mathbf{x} = \mathbf{z}_0 + \mathbf{z}_1$. Then,

$$\| \mathbf{x} - \mathbf{x} \|^2 = 0,$$

or equivalently,

$$\| (\mathbf{x}_0 + \mathbf{x}_1) - (\mathbf{z}_0 + \mathbf{z}_1) \|^2 = 0.$$

After rearranging terms this equation becomes,

$$\| (\mathbf{x}_0 - \mathbf{z}_0) + (\mathbf{x}_1 - \mathbf{z}_1) \|^2 = 0.$$

Then, using the Pythagorean Theorem,

$$\| \mathbf{x}_0 - \mathbf{z}_0 \|^2 + \| \mathbf{x}_1 - \mathbf{z}_1 \|^2 = 0.$$

Therefore, $\mathbf{z}_0 = \mathbf{x}_0$ and $\mathbf{z}_1 = \mathbf{x}_1$. Hence, the decomposition is unique. ∎

5.3.1.5 Multiresolution analysis

Definition 5.3.1.6. *Multiresolution analysis (MRA) can be viewed as a sequence of approximations of a given function $f(t)$ at different resolutions. The approximation of $f(t)$ at a resolution 2^j is defined as an orthogonal projection of $f(t)$ on a subspace $\mathbf{V}_j$. Now, we will provide a list of properties that these subspaces will need to satisfy. They are:*

1. $\mathbf{V}_j \subset \mathbf{V}_{j+1}$;
2. $\cap_j \mathbf{V}_j = \{\mathbf{0}\}$ *and closure of* $\cup_j \mathbf{V}_j = \mathbf{L}^2$;
3. $f \in \mathbf{V}_{j+1} \Leftrightarrow f(2\bullet) \in \mathbf{V}_j$;
4. $f \in \mathbf{V}_j \Rightarrow f(\bullet - k) \in \mathbf{V}_j$ *for all integer k*;
5. $\{f(\bullet - k)\}$ *is a Reisz basis in* $\mathbf{V}_0$.

A level of a multiresolution analysis is one of the $\mathbf{V}_j$ subspaces and one level is coarser (finer) with respect to another whenever the index of the corresponding subspace is smaller (bigger). By properties 3, 4, and 5 there exists a Reisz basis φ in $\mathbf{V}_j$ of the form

$$\varphi_{j,k}(t) = c\varphi \left(2^j t - k \right)$$

where, c is a constant. Assume that the energy associated with $\varphi(t)$ equals the energy associated with $\varphi_{j,k}(t)$, that is,

$$\int_{-\infty}^{\infty} \varphi_{j,k}^*(t)\varphi_{j,k}(t)dt = \int_{-\infty}^{\infty} \varphi^*(t)\varphi(t)dt,$$

or equivalently,

$$c^2 \int_{-\infty}^{\infty} \varphi^* \left(2^j t - k\right) \varphi \left(2^j t - k\right) dt = \int_{-\infty}^{\infty} \varphi^*(t)\varphi(t)dt.$$

For the left-hand side, perform a change of variables by letting $u = 2^j t - k$. Then,

$$\frac{c^2}{2^j} \int_{-\infty}^{\infty} \varphi^*(u)\varphi(u)du = \int_{-\infty}^{\infty} \varphi^*(t)\varphi(t)dt.$$

Therefore,

$$c = 2^{j/2}.$$

Hence, the scaling function for space $\mathbf{V}_j$ has the form

$$\varphi_{j,k}(t) = 2^{j/2}\varphi \left(2^j t - k\right).$$

In an analogous fashion, the wavelet function for the subspace $\mathbf{W}_j$, which is the complement space of $\mathbf{V}_j$, has the form

$$\psi_{j,k}(t) = 2^{j/2}\psi \left(2^j t - k\right).$$

Since information about resolution 2^{j+1} is given by the *approximation* in subspace $\mathbf{V}_j$ and the *details* in subspace $\mathbf{W}_j$, the orthogonal projection on the orthogonal complement of space $\mathbf{V}_j$ in space $\mathbf{V}_{j+1}$, that is

$$\mathbf{V}_{j+1} = \mathbf{V}_j \oplus \mathbf{W}_j$$

where,

$$f_j \in \mathbf{V}_j \text{ and } (f_{j+1} - f_j) \in \mathbf{W}_j.$$

Since $\mathbf{V}_1 = \mathbf{V}_0 \oplus \mathbf{W}_0$ and $\mathbf{V}_2 = \mathbf{V}_1 \oplus \mathbf{W}_1$ then, $\mathbf{V}_2 = \mathbf{V}_0 \oplus [\mathbf{W}_0 \oplus \mathbf{W}_1]$. So, in general,

$$\mathbf{V}_{j+1} = \mathbf{V}_0 \oplus [\mathbf{W}_0 \oplus \mathbf{W}_1 \oplus \cdots \oplus \mathbf{W}_j],$$

or simply,

$$\mathbf{V}_{j+1} = \mathbf{V}_0 \oplus_{0 \leq k \leq j} \mathbf{W}_k$$

where, $\mathbf{V}_0$ is the subspace corresponding to the coarsest scale. Since $\varphi \in \mathbf{V}_0$ and $\{\psi_{j,k}\} \in \oplus_{0 \leq k \leq j} \mathbf{W}_k$, then, an input signal $f(x)$ can be expressed as

$$f(x) = s\varphi(x) + \sum_j \sum_k d_{j,k}\psi_{j,k}(x)$$

where,

$$s = \langle f, \varphi \rangle = \int_{-\infty}^{\infty} \varphi(x)f(x)dx$$

and

$$d_{j,k} = \langle f, \psi_{j,k} \rangle = \int_{-\infty}^{\infty} \psi_{j,k}(x)f(x)dx = 2^{j/2} \int_{-\infty}^{\infty} \psi(2^j x - k)f(x)dx.$$

Let us consider the following example to illustrate many of the ideas of multiresolution analysis. Assume the input signal $f(x)$ is defined by

$$f(x) = \begin{cases} 16, & 0 \le x < \frac{1}{4} \\ -8, & \frac{1}{4} \le x < \frac{1}{2} \\ 16, & \frac{1}{2} \le x < \frac{3}{4} \\ -20, & \frac{3}{4} \le x < 1 \\ 0, & \text{otherwise}. \end{cases}$$

For simplicity of this example, we will use the Haar wavelet. Its wavelet, $\psi(x)$, and scaling function, $\varphi(x)$, are defined by

$$\varphi(x) = \begin{cases} 1, & 0 \le x < 1 \\ 0, & \text{otherwise} \end{cases}$$

and

$$\psi(x) = \begin{cases} 1, & 0 \le x < \frac{1}{2} \\ -1, & \frac{1}{2} \le x < 1 \\ 0, & \text{otherwise}. \end{cases}$$

Now, let us compute the coefficients.

$$\begin{aligned} s \quad &= \quad \langle f, \varphi \rangle = \int_{-\infty}^{\infty} \varphi(x)f(x)dx \\ &= \quad (1)(16)(\tfrac{1}{4}) + (1)(-8)(\tfrac{1}{4}) + (1)(16)(\tfrac{1}{4}) + (1)(-20)(\tfrac{1}{4}) = 1, \end{aligned}$$

$$\begin{aligned} d_{0,0} \quad &= \quad \langle f, \psi_{0,0} \rangle = \int_{-\infty}^{\infty} \psi_{0,0}(x)f(x)dx = \int_{-\infty}^{\infty} \psi(x)f(x)dx \\ &= \quad (1)(16)(\tfrac{1}{4}) + (1)(-8)(\tfrac{1}{4}) + (-1)(16)(\tfrac{1}{4}) + (-1)(-20)(\tfrac{1}{4}) = 3, \end{aligned}$$

$$\begin{aligned} d_{1,0} \quad &= \quad \langle f, \psi_{1,0} \rangle = \int_{-\infty}^{\infty} \psi_{1,0}(x)f(x)dx = \sqrt{2} \int_{-\infty}^{\infty} \psi(2x)f(x)dx \\ &= \quad (\sqrt{2})(16)(\tfrac{1}{4}) + (-\sqrt{2})(-8)(\tfrac{1}{4}) = 6\sqrt{2}, \end{aligned}$$

$$\begin{aligned} d_{1,1} \quad &= \quad \langle f, \psi_{1,1} \rangle = \int_{-\infty}^{\infty} \psi_{1,1}(x)f(x)dx = \sqrt{2} \int_{-\infty}^{\infty} \psi(2x - 1)f(x)dx \\ &= \quad (\sqrt{2})(16)(\tfrac{1}{4}) + (-\sqrt{2})(-20)(\tfrac{1}{4}) = 9\sqrt{2}. \end{aligned}$$

Hence,

$$
\begin{bmatrix} 16 \\ -8 \\ 16 \\ -20 \end{bmatrix} = \begin{bmatrix} 1 \\ 1 \\ 1 \\ 1 \end{bmatrix} + 3 \begin{bmatrix} 1 \\ 1 \\ -1 \\ -1 \end{bmatrix} + 6\sqrt{2} \begin{bmatrix} \sqrt{2} \\ -\sqrt{2} \\ 0 \\ 0 \end{bmatrix} + 9\sqrt{2} \begin{bmatrix} 0 \\ 0 \\ \sqrt{2} \\ -\sqrt{2} \end{bmatrix}.
$$

Thus,

$$
\underbrace{\begin{bmatrix} 16 \\ -8 \\ 16 \\ -20 \end{bmatrix}}_{\mathbf{y}} = \underbrace{\begin{bmatrix} 1 & 1 & \sqrt{2} & 0 \\ 1 & 1 & -\sqrt{2} & 0 \\ 1 & -1 & 0 & \sqrt{2} \\ 1 & -1 & 0 & -\sqrt{2} \end{bmatrix}}_{\mathbf{W_4}} \underbrace{\begin{bmatrix} 1 \\ 3 \\ 6\sqrt{2} \\ 9\sqrt{2} \end{bmatrix}}_{\mathbf{b}}
$$

where,

$$
\mathbf{b} = [s, \ d_{0,0}, \ d_{1,0}, \ d_{1,1}]^T
$$

and $\mathbf{y}$ is a vector of input signal values, and $\mathbf{W_4}$ is the discrete Haar wavelet transform. The discrete wavelet transform matrix is analogous to the discrete Fourier transform matrix — the principal difference being that Fourier coefficients come from values at n points, while Haar wavelet coefficients come from n subintervals.

5.3.1.6 Scaling functions

According to the definition of the multiresolution analysis, $\varphi \in \mathbf{V}_0 \subset \mathbf{V}_1$. As such, the scaling function must satisfy

$$
\varphi(t) = \sqrt{2} \sum_k h_0(k)\varphi(2t - k),
$$

or equivalently,

$$
\varphi_{0,0}(t) = \sum_k h_0(k)\varphi_{1,k}(t).
$$

This equation is known by several different names: the refinement equation, the dilation equation, and the two-scale difference equation. The generalization of the refinement equation is given by

$$
\varphi_{j,n}(t) = \sum_k h_0(k + 2n)\varphi_{j+1,k}(t).
$$

The following theorem provides an important property of the sequence of filter (or mask) coefficients $\{h_0\}$.

Theorem 5.3.1.5. $\sum_k h_0(k) = \sqrt{2}$.

Proof: Integrating both sides of the refinement equation yields

$$\int_{-\infty}^{\infty} \varphi(t)dt = \int_{-\infty}^{\infty} \sqrt{2} \sum_{k} h_0(k)\varphi(2t-k)dt,$$

or equivalently,

$$\int_{-\infty}^{\infty} \varphi(t)dt = \sqrt{2} \sum_{k} h_0(k) \int_{-\infty}^{\infty} \varphi(2t-k)dt.$$

For the integral on the right-hand-side, perform a change of variables by letting $u = 2t - k$. Then,

$$\int_{-\infty}^{\infty} \varphi(t)dt = \frac{1}{\sqrt{2}} \sum_{k} h_0(k) \int_{-\infty}^{\infty} \varphi(u)du.$$

Hence,

$$\sum_{k} h_0(k) = \sqrt{2}.$$

∎

The scaling function is uniquely defined by the refinement equation and the normalization

$$\int_{-\infty}^{\infty} \varphi(t)dt = 1.$$

5.3.1.7   Wavelets

Since $\mathbf{V}_1 = \mathbf{V}_0 \oplus \mathbf{W}_0$, then $\psi \in \mathbf{V}_1$. As such, the wavelet must satisfy

$$\psi(t) = \sqrt{2} \sum_{k} h_1(k)\varphi(2t-k),$$

or equivalently,

$$\psi_{0,0}(t) = \sum_{k} h_1(k)\varphi_{1,k}(t).$$

The generalization of this equation is given by

$$\psi_{j,n}(t) = \sum_{k} h_1(k+2n)\varphi_{j+1,k}(t).$$

The wavelet is uniquely defined by this equation and by the normalization

$$\int_{-\infty}^{\infty} \psi(t)dt = 0.$$

The following theorem provides an important property of the sequence of coefficients $\{h_1\}$.

Theorem 5.3.1.6. $\sum_{k} h_1(k) = 0$.

Proof: Since the wavelet ψ must satisfy

$$\psi(t) = \sqrt{2} \sum_k h_1(k)\varphi(2t - k),$$

then by integrating both sides of the equation yields

$$\int_{-\infty}^{\infty} \psi(t)dt = \int_{-\infty}^{\infty} \sqrt{2} \sum_k h_1(k)\varphi(2t - k)dt,$$

or equivalently,

$$\int_{-\infty}^{\infty} \psi(t)dt = \sqrt{2} \sum_k h_1(k) \int_{-\infty}^{\infty} \varphi(2t - k)dt.$$

For the integral on the right-hand side, perform a change of variables by letting $u = 2t - k$. Then,

$$\int_{-\infty}^{\infty} \psi(t)dt = \frac{1}{\sqrt{2}} \sum_k h_1(k) \int_{-\infty}^{\infty} \varphi(u)du.$$

Since $\int_{-\infty}^{\infty} \psi(t)dt = 0$ and $\int_{-\infty}^{\infty} \varphi(u)du = 1$, then

$$\sum_k h_1(k) = 0.$$

■

5.3.1.8 Filter-bank implementation

The projection of a signal $f(t) \in \mathbf{V}_{j+1}$ is given by

$$f(t) = \sum_n a_{j+1}(n)\varphi_{j+1,n}(t)$$

where,

$$a_{j+1}(n) = \langle f, \varphi_{j+1,n} \rangle .$$

In addition, $f(t) \in \mathbf{V}_j \oplus \mathbf{W}_j$, so $f(t)$ can also be written as

$$f(t) = \sum_k a_j(k)\varphi_{j,k}(t) + \sum_l b_j(l)\psi_{j,l}(t)$$

where

$$a_j(k) = \langle f, \varphi_{j,k} \rangle \text{ and } b_j(l) = \langle f, \psi_{j,l} \rangle .$$

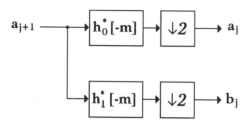

Figure 5.1. An analysis stage of pyramidal filter bank.

Let us examine the coefficients a little closer.

$$
\begin{aligned}
a_j(k) &= \langle f, \varphi_{j,k} \rangle \\
&= \langle f, \sum_m h_0(2k+m) \; \varphi_{j+1,m} \rangle \\
&= \sum_m h_0^*(2k+m) \langle f, \varphi_{j+1,m} \rangle \\
&= \sum_m h_0^*(2k+m) a_{j+1}(m)
\end{aligned}
$$

and

$$
\begin{aligned}
b_j(k) &= \langle f, \psi_{j,k} \rangle \\
&= \langle f, \sum_m h_1(2k+m) \; \varphi_{j+1,m} \rangle \\
&= \sum_m h_1^*(2k+m) \langle f, \; \varphi_{j+1,m} \rangle \\
&= \sum_m h_1^*(2k+m) a_{j+1}(m).
\end{aligned}
$$

These equations suggest a pyramidal filter bank implementation, where the analysis bank is defined in terms of stages, where $a_{j+1}(m) = \langle f, \; \varphi_{j+1,m} \rangle$ is decomposed into $a_j(k) = \langle f, \varphi_{j,k} \rangle$ and $b_j(k) = \langle f, \psi_{j,k} \rangle$. This result can be depicted pictorially in Figure 5.1. Thus, we have defined the analysis side of the filter bank. Now, let us examine the synthesis side.

Since $V_{j+1} = V_j \oplus W_j$, we can write any vector in V_{j+1} as a linear combination of the basis vectors in V_j and W_j. Since $V_j \subset V_{j+1}$, a shift of n in V_{j+1} corresponds to a shift of $\frac{n}{2}$ in V_j and W_j. That is,

$$
\varphi_{j+1,n}(t) = \sum_k q_0(k) \varphi_{j,k+\frac{n}{2}}(t) \; + \; \sum_l p_0(l) \psi_{j,l+\frac{n}{2}}(t).
$$

Let $g_0(2k) = q_0(k)$ and $g_1(2l) = p_0(l)$. Then,

$$
\varphi_{j+1,n}(t) = \sum_k g_0(2k) \varphi_{j,k+\frac{n}{2}}(t) \; + \; \sum_l g_1(2l) \psi_{j,l+\frac{n}{2}}(t).
$$

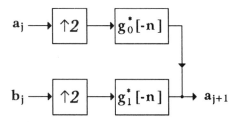

Figure 5.2. A synthesis stage of pyramidal filter bank.

Let $m_0 = k + \frac{n}{2}$ and $m_1 = l + \frac{n}{2}$. Then,

$$\varphi_{j+1,n}(t) = \sum_{m_0} g_0(2m_0 - n)\varphi_{j,m_0}(t) + \sum_{m_1} g_1(2m_1 - n)\psi_{j,m_1}(t).$$

Then, the projection of signal $f(t)$ on the basis elements $\varphi_{j+1,n}(t)$ is given by

$$
\begin{aligned}
a_{j+1}(n) &= \langle f, \varphi_{j+1,n} \rangle \\
&= \left\langle f, \sum_{m_0} g_0(2m_0 - n)\varphi_{j,m_0} + \sum_{m_1} g_1(2m_1 - n)\psi_{j,m_1} \right\rangle \\
&= \left\langle f, \sum_{m_0} g_0(2m_0 - n)\varphi_{j,m_0} \right\rangle + \left\langle f, \sum_{m_1} g_1(2m_1 - n)\psi_{j,m_1} \right\rangle \\
&= \sum_{m_0} g_0^*(2m_0 - n) \langle f, \varphi_{j,m_0} \rangle + \sum_{m_1} g_1^*(2m_1 - n) \langle f, \psi_{j,m_1} \rangle \\
&= \sum_{m_0} g_0^*(2m_0 - n) \, a_j(m_0) + \sum_{m_1} g_1^*(2m_1 - n) \, b_j(m_1).
\end{aligned}
$$

This equation suggests a pyramidal filter-bank implementation where the synthesis bank is defined in terms of stages, where $a_j(m_0) = \langle f, \varphi_{j,m_0} \rangle$ and $b_j(m_1) = \langle f, \psi_{j,m_1} \rangle$ are merged to form $a_{j+1}(n) = \langle f, \varphi_{j+1,n} \rangle$. This result is depicted pictorially in Figure 5.2.

5.3.1.9 Initial projection coefficients

The accuracy of the multiresolution analysis depends on the computation of $a_N(k)$, *i.e.*,

$$a_N(k) = \langle f, \varphi_{N,k} \rangle = 2^{N/2} \int_{-\infty}^{\infty} f(t) \, \varphi(2^N t - k) \, dt.$$

Let us perform a change of variables by letting $u = 2^N t$. Then,

$$a_N(k) = 2^{-N/2} \int_{-\infty}^{\infty} f(2^{-N} u) \, \varphi(u - k) \, du.$$

Since $\varphi(u)$ has compact support, this integral can be interpreted to have finite limits. In addition, $f(2^{-N}u)$ may place constraints on the sampling of $f(u)$, depending of course on the smoothness of $f(u)$.

5.3.1.10   Reflections on the wavelet transform

By examining multiresolution analysis, we are ready to draw some interesting conclusions about wavelets.

1. *Concept of Scale.* The discrete wavelet transform is useful for representing the finer *variations* in the signal $f(t)$ at various scales. Moreover, the function $f(t)$ can be represented as a linear combination of functions that represent the variations at different scales.

2. *Localized Basis.* If a signal has energy at a particular scale concentrated in an interval in the time domain, then the corresponding coefficient has a large value. Therefore, the wavelet basis provides localization information in both the time domain as well as the scale domain.

5.3.2   Image compression and wavelets

In multiresolution analysis, there are some dependencies between coefficients at different scales. In addition, for natural images, it is unlikely to find significant high-frequency energy if there is little low-frequency energy in the same spatial location. These observations motivated a wavelet-based image coding method, that is based on a data structure called a zero-tree.

A zero-tree provides a compact representation to indicate the positions of *significant* wavelet coefficients. A wavelet coefficient c is *insignificant* with respect to a threshold T, if $|c| < T$. The zero-tree is based on the hypothesis that if a wavelet coefficient at a given scale is insignificant with respect to a given threshold T, then, in most cases, all wavelet coefficients at the same orientation in the same spatial location at finer scales are insignificant with respect to T. An engineering trade-off can be made between the choice of the threshold T and the number of bits of precision for unthresholded coefficients.

Thus, the zero-tree approach can isolate interesting nonzero details by eliminating large insignificant regions from consideration; and, therefore, it makes more bits available to encode the significant coefficients.

5.3.3   A generalization: biorthogonal wavelets

Except for the Haar wavelet, all other orthogonal wavelets are not symmetric. This is a problem in applications, such as image compression, since these unsymmetrical wavelets generate noticeable phase distortions. So can orthogonal wavelets be generalized to eliminate this problem? Yes, using biorthogonal wavelets.

We will develop this idea of biorthogonal wavelets by first imagining two intertwining multiresolution analyses, that is,

$$\mathbf{V}_{j+1} = \mathbf{V}_j \oplus \mathbf{W}_j \text{ and } \widetilde{\mathbf{V}}_{j+1} = \widetilde{\mathbf{V}}_j \oplus \widetilde{\mathbf{W}}_j$$

where,

$$\mathbf{V}_j \cap \mathbf{W}_j = 0 \text{ and } \widetilde{\mathbf{V}}_j \cap \widetilde{\mathbf{W}}_j = 0.$$

Instead of assuming the spaces $\mathbf{V}_j$ and $\mathbf{W}_j$ are perpendicular, we will assume that $\mathbf{V}_j$ is perpendicular to $\widetilde{\mathbf{W}}_j$. Similarly, we will assume that the spaces $\widetilde{\mathbf{V}}_j$ and $\widetilde{\mathbf{W}}_j$ are not perpendicular, but rather that $\widetilde{\mathbf{V}}_j$ is perpendicular to $\mathbf{W}_j$. Then, the wavelets $\psi_{j,k}(t) = 2^{j/2}\psi(2^j t - k)$ and $\widetilde{\psi}_{l,m}(t) = 2^{l/2}\widetilde{\psi}(2^l t - m)$ are biorthogonal wavelet bases. That is, the wavelet bases satisfy the following biorthonormality condition:

$$\int_{-\infty}^{\infty} \psi_{j,k}(t)\widetilde{\psi}_{l,m}(t)dt = \delta_{j,l}\delta_{k,m}.$$

In addition, the scaling functions $\varphi_{j,k}(t) = 2^{j/2}\varphi(2^j t - k)$ and $\widetilde{\varphi}_{j,m}(t) = 2^{j/2}\widetilde{\varphi}(2^j t - m)$ satisfy the following orthonormality condition:

$$\int_{-\infty}^{\infty} \varphi_{j,k}(t)\widetilde{\varphi}_{j,m}(t)dt = \delta_{k,m}.$$

5.3.4 Case study: wavelets in data communications

Although this case study is focused on data communications, many of the ideas expressed are also appropriate for other applications. Now, let us consider a modulated signal, denoted $x(t)$, satisfying the dyadic self-similarity property is given by

$$x(t) = 2^{-mH}x(2^m t)$$

for some real constant H and all integers m. This is the class of signals of interest for *fractal modulation* applications. These waveforms have the property that an arbitrary short duration time segment is sufficient to recover the entire waveform and hence the embedded information. Similarly, a low bandwidth approximation to the waveform is sufficient to recover the undistorted waveform and also the embedded information.

Now, consider the wavelet representation of $x(t)$, that is,

$$x(t) = \sum_m \sum_n x_{m,n}\psi_{m,n}(t)$$

where,

$$x_{m,n} = \langle x, \psi_{m,n} \rangle = 2^{m/2} \int_{-\infty}^{\infty} x(t)\psi(2^m t - n)dt.$$

Let us examine further $x_{m,n}$. Substituting the definition of $x(t)$ into $x_{m,n}$ yields

$$x_{m,n} = 2^{m/2} \int_{-\infty}^{\infty} 2^{-mH} x(2^m t)\psi(2^m t - n)dt.$$

Perform a change of variables by letting $u = 2^m t$. Then,

$$x_{m,n} = 2^{-m(2H+1)/2} \int_{-\infty}^{\infty} x(u)\psi(u - n)du$$

or equivalently,

$$x_{m,n} = \beta^{-m/2} x_{o,n}$$

where $\beta = 2^{2H+1}$. Let us denote $x_{o,n}$ by $q(n)$. Then,

$$x(t) = \sum_m \sum_n \beta^{-m/2} q(n)\psi_{m,n}(t).$$

We wish to let $q(n)$ correspond to a finite length message of length L that we wish to send, but the right-hand side of this equation is an infinite sum. Let's use periodic extensions of $q(n)$ in $x(t)$, that is,

$$x(t) = \sum_m \sum_n \beta^{-m/2} q(n \bmod L)\psi_{m,n}(t).$$

Moreover, since $\beta = 2^{2H+1}$, the parameter H controls the relative power distribution among the frequency bands and hence the overall transmitted power. Assume that $x(t)$ is transmitted across the channel and received as a waveform denoted $r(t)$. Then, the receiver extracts the wavelet coefficients $r_{m,n}$ of $r(t)$ using the discrete wavelet transform. These coefficients take the form

$$r_{m,n} = \beta^{-m/2} q(n \bmod L) + z_{m,n}$$

where $z_{m,n}$ are the wavelet coefficients of the channel noise.

The duration-bandwidth characteristics of the channel affect which observation coefficients $r_{m,n}$ may be accessed. If the channel is bandlimited to 2^{M_u} Hz for some integer M_u, then the receiver can not access coefficients at scales $m > M_u$. Similarly, the duration constraint of the channel results in a minimum decoding rate of 2^{M_L} symbols/second for some integer M_L, which precludes access to scales $m < M_L$.

Therefore, the coefficients available at the receiver correspond to the indices

$$m = M_L, M_L + 1, \ldots, M_U$$

and

$$n = 0, 1, \ldots, L2^{m-M_L} - 1$$

where L is the length of the message $q(n)$. Therefore, the number of noisy measurements of the message at the receiver is given by

$$K = \sum_{m=M_L}^{M_U} 2^{m-M_L}.$$

Perform a change of variables by letting $p = m - M_L$ yields

$$K = \sum_{p=0}^{M_U-M_L} 2^P = 2^{M_U-M_L+1} - 1.$$

Thus, the receiver can select the rate/bandwidth ratio dynamically since the transmitter is fixed.

5.4   Problems

1. Let

$$F(\omega) = \sum_{k=-\infty}^{\infty} f(k)\eta_k(\omega)$$

where $\eta_k(\omega)$ is a set of orthonormal functions in the range $a \leq \omega \leq b$, that is,

$$\int_a^b \eta_k(\omega)\eta_p^*(\omega)d\omega = \delta_{k,p}.$$

Suppose we wish to approximate $F(\omega)$ with the finite summation

$$F_m(\omega) = \sum_{p=-m}^{m} \widehat{f}(p)\eta_p(\omega)$$

where $a \leq \omega \leq b$. We wish to make the approximation best in the least square sense, that is

$$\text{error} = \int_a^b |F(\omega) - F_m(\omega)|^2 d\omega$$

must be minimized. Show that the choice of $\widehat{f}(k) = f(k)$, $k = -m, \ldots, m$, achieves the desired result.

2. Let us examine the Mexican hat function, which is defined by

$$h(t) = (1 - t^2)\exp(-t^2/2).$$

Is $h(t)$ a wavelet? Justify your answer.

3. Given n-dimensional vectors $\mathbf{s}$ and $\mathbf{t}$ and an $n \times n$ matrix $\mathbf{W}$. Show that $\mathbf{t}^H\mathbf{W}\mathbf{s}$ satisfies the properties of an inner product, $i.e.$

$$\langle \mathbf{s}, \mathbf{t} \rangle = \mathbf{t}^H\mathbf{W}\mathbf{s}.$$

4. Prove the following: Let $y_{j,k}(t) = 2^{j/2}y(2^jt - k)$. If $y(t)$ has unit norm, then $y_{j,k}(t)$ also has unit norm.

Bibliography

[1] P. Auscher, G. Weiss, and M. V. Wickerhauser, Local sine and cosine basis of Coifman and Meyer and the construction of smooth wavelets, *Wavelets: A Tutorial in Theory and Applications* (C. K. Chui, ed.), Academic Press, Boston, 1992, pp. 237–256.

[2] V. Belevitch, *Classical Network Theory*, Holden Day, Inc., San Francisco, 1968.

[3] M. Bellanger, G. Bonnerot, and M. Condreuse, Digital filtering by polyphase network: Application to sampling rate alteration and filter banks, *IEEE Trans. on Acoust., Speech, and Signal Processing* **24** (1976), 109–114.

[4] J. W. S. Cassels, *An Introduction to the Geometry of Numbers*, Springer-Verlag, Berlin, Germany, 1959.

[5] Y. T. Chan, *An Introduction to Wavelets*, Kluwer Academic Publishers, Boston, 1992.

[6] T. Chen and P. P. Vaidyanathan, Recent developments in multidimensional multirate systems, *IEEE Trans. on Circuits and Syst. for Video Tech.* **3** (1993), 116–137.

[7] C. K. Chui, *An Introduction to Wavelets*, Academic Press, Boston, 1992.

[8] A. Cohen, I. Daubechies, and J. C. Feauveau, Biorthogonal bases of compactly supported wavelets, *Comm. Pure Appl. Math.* **45** (1992), 485–560.

[9] R. Coifman and Y. Meyer, Remarques sur l'analyse de fourier à fenêtre, *Comptes Rendus Acad. Sci. Paris Série I* **312** (1991), 259–261.

[10] R. E. Crochiere and L. R. Rabiner, Interpolation and decimation of digital signals: A tutorial review, *Proc. of IEEE* **69** (1981), 300–331.

[11] ———, *Multirate Digital Signal Processing*, Prentice Hall, Englewood Cliffs, NJ, 1983.

[12] I. Daubechies, Orthonormal bases of compactly supported wavelets, *Comm. Pure Appl. Math.* **41** (1988), 909–996.

[13] ———, *Ten Lectures on Wavelets*, SIAM, Philadelphia, 1992.

[14] Z. Doğanata and P. P. Vaidyanathan, Minimal structures for the implementation of digital rational lossless systems, *IEEE Trans. on Acoustics, Speech, and Signal Processing* **38** (1990), 2058–2074.

[15] E. Dubois, The sampling and reconstruction of time-varying imagery with applications in video systems, *Proc. of IEEE* **73** (1985), 502–522.

[16] D. E. Dudgeon and R. M. Mersereau, *Multidimensional Digital Signal Processing*, Prentice-Hall, Englewood-Cliffs, NJ, 1984.

[17] G. Evangelista, Comb and multiplexed wavelet transforms and their application to signal processing, *IEEE Trans. on Signal Processing* **42** (1994), 292–303.

[18] B. L. Evans, R. H. Bamberger, and J. H. McClellan, Rules for multidimensional multirate structures, *IEEE Trans. on Signal Processing* **42** (1994), 762–771.

[19] N. J. Fliege, *Multirate Digital Signal Processing*, John Wiley, Chichester, England, 1994.

[20] A. Guessoum and R. M. Mersereau, Fast algorithms for the multidimensional discrete Fourier transform, *IEEE Trans. on Acoustics, Speech, and Signal Processing* **34** (1986), 937–943.

[21] E. Gündüzhan, A. E. Cetin, and A. M. Tekalp, DCT coding of nonrectangularly sampled images, *IEEE Signal Processing Letters* **9** (1994), 131–133.

[22] M. Holschneider, *Wavelets: An Analysis Tool*, Oxford Science Publications, Clarendon Press, Oxford, England, 1995.

[23] G. Karlsson and M. Vetterli, Theory of two-dimensional multirate filter banks, *IEEE Trans. on Acoustics, Speech, and Signal Processing* **38** (1990), 925–937.

[24] R. D. Koilpillai and P. P. Vaidyanathan, Cosine-modulated FIR filter banks satisfying perfect reconstruction, *IEEE Trans. on Signal Processing* **40** (1992), 770–783.

[25] J. Kovačević, Subband coding systems incorporating quantizer models, *IEEE Trans. on Image Processing* **4** (1995), 543–553.

[26] J. S. Lim, *Two-Dimensional Signal and Image Processing*, Prentice-Hall, Englewood Cliffs, NJ, 1990.

[27] S. Mallat, Multiresolution approxmations and wavelet orthoonormal bases, *Trans. of Amer. Math. Soc.* **11** (1989), 64–88.

[28] ———, A theory of multiresolution signal decomposition: The wavelet representation, *IEEE Trans. on Pattern Analysis and Machine Intelligence* **11** (1989), 674–693.

[29] H. S. Malvar, Theory of two-dimensional multirate filter banks, *IEEE Trans. on Acoustics, Speech, and Signal Processing* **38** (1990), 969–978.

[30] ———, *Signal Processing with Lapped Transforms*, Artech House, Norwood, MA, 1992.

[31] R. M. Mersereau and T. C. Speake, The processing of periodically sampled multidimensional signals, *IEEE Trans. on Acoustics, Speech, and Signal Processing* **31** (1983), 188–194.

[32] R. A. Meyer and C. S. Burrus, A unified analysis of multirate and periodically time-varying digital filters, *IEEE Trans. on Circuits and Syst.* **22** (1975), 162–168.

[33] Y. Meyer, *Wavelets: Algorithms and Applications*, SIAM, Philadelphia, 1993.

[34] ———, *Wavelets and Operators*, Cambridge University Press, Cambridge, England, 1993.

[35] F. Mintzer, Filters for distortion-free two-band multirate filter banks, *IEEE Trans. on Acoustics, Speech, and Signal Processing* **33** (1985), 626–630.

[36] J. Morlet, G. Arens, I. Fourgeau, and D. Giard, Wave propagation and sampling theory., *Geophysics* **47** (1982), 203–236.

[37] G. Oetken, T. W. Parks, and W. Schüssler, New results on the design of digital interpolators, *IEEE Trans. on Acoustics, Speech, and Signal Processing* **23** (1975), 301–309.

[38] D. P. Petersen and D. Middleton, Sampling and reconstruction of wave-number-limited functions in N-dimensional euclidean space, *Information and Control* **5** (1962), 279–323.

[39] F. Riesz and B. Sz.-Nagy, *Functional Analysis*, Frederick Ungar Publishing Co., New York, 1955.

[40] R. W. Schafer and L. R. Rabiner, A digital signal processing approach to interpolation, *Proc. of IEEE* **61** (1973), 692–702.

[41] R. Schilling and H. Lee, *Engineering Analysis: A Vector Space Approach*, John Wiley, New York, 1988.

[42] J. M. Shapiro, Embedded image coding using zerotrees of wavelet coefficients, *IEEE Trans. on Signal Processing* **41** (1993), 3445–3462.

[43] H. J. S. Smith, On systems of linear indeterminate equations and congruences, *Phil. Trans. Royal Soc. London* **151** (1861), 293–326.

[44] M. J. T. Smith and T. P. Barnwell III, Exact reconstruction from tree-structured subband coders, *IEEE Trans. on Acoustics, Speech, and Signal Processing* **34** (1986), 431–441.

[45] _____, A new filter bank theory for time-frequency representations, *IEEE Trans. on Acoustics, Speech, and Signal Processing* **35** (1987), 314–327.

[46] G. Strang and T. Q. Nguyen, *Wavelets and Filter Banks*, Wellesley-Cambridge Press, Wellesley, MA, 1996.

[47] B. W. Suter and M. E. Oxley, On variable overlapped windows and weighted orthonormal bases, *IEEE Trans. on Signal Processing* **42** (1994), 1973–1982.

[48] P. P. Vaidyanathan, Theory and design of M-channel maximally decimated quadrature mirror filters with arbitrary M, having the perfect reconstruction property., *IEEE Trans. on Acoustics, Speech, and Signal Processing* **35** (1987), 476–492.

[49] _____, *Multirate Systems and Filter Banks*, Prentice-Hall, Englewood-Cliffs, NJ, 1993.

[50] P. P. Vaidyanathan and S. K. Mitra, Polyphase networks, block digital filtering, LPTV systems, and alias-free QMF banks: A unified approach based on pseudocirculants, *IEEE Trans. on Acoustics, Speech, and Signal Processing* **36** (1988), 381–391.

[51] P. P. Vaidyanathan, T. Q. Nguyen, Z. Doğanata, and T. Saramäki, Improved technique for design of perfect reconstruction FIR QMF banks with lossless polyphase matrices, *IEEE Trans. on Acoustics, Speech, and Signal Processing* **37** (1989), 1042–1056.

[52] M. Vetterli, Multidimensional subband coding: Some theory and algorithms, *Signal Processing* **6** (1984), 97–112.

[53] M Vetterli, Filter banks allowing for perfect reconstruction, *Signal Processing* **10** (1986), 219–244.

[54] M. Vetterli and J. Kovačević, *Wavelets and Subband Coding*, Prentice-Hall, Englewood Cliffs, NJ, 1995.

[55] E. Viscito and J. P. Allebach, The analysis and design of multidimensional FIR perfect reconstruction filter banks, *IEEE Trans. on Circuits and Syst.* **38** (1991), 29–41.

[56] P. H. Westerink, J. Biemond, and D. E. Boekee, Scalar quantization error analysis for image subband coding using QMFs, *IEEE Trans. on Signal Processing* **40** (1992), 421–428.

[57] G. W. Wornell, *Signal Processing with Fractals: A Wavelet-Based Approach*, Prentice Hall, Upper Saddle River, NJ, 1996.

[58] X.-G. Xia and B. W. Suter, A family of two-dimensional nonseparable malvar wavelets, *Applied and Computational Harmonic Analysis* **2** (1995), 243–256.

[59] ———, Construction of malvar wavelets on hexagons, *Applied and Computational Harmonic Analysis* **3** (1996), 65–71.

Index

197

WAVELET ANALYSIS AND ITS APPLICATIONS

CHARLES K. CHUI, SERIES EDITOR